1

Universities Under Fire

The culture wars have well and truly breached the campus walls. Emboldened by recent electoral success, the political right around the world has taken a chainsaw to higher education. It is a tale as old as time. Wherever conservatism rears its head, universities are first in the firing line. Academia has long served as a canary in the mine of demagoguery.

Despotism is always quick to detect, and then to reject, critical scrutiny. Ever adept at exploiting the natural human wariness of the unfamiliar, it also thrives on scapegoating the esoteric. Enraging the masses is a form of misdirection long favoured by authoritarians. When Hungary banned its universities from teaching gender studies in 2018, it did so on the pretext that such a topic was 'destructive' to Hungarian moral integrity (Stanley, 2018), thereby depicting a field long influenced by Freud's uniquely Austro-Hungarian approach to psychoanalysis as nothing less than a threat to human society. The erosion of democracy is easier to conceal when morality itself needs protecting.

Attacks on academia serve power by pushing back on those who would wish to dismantle power hierarchies. Often this involves the perverting of narratives, where aspirations for justice are cast as anything but. And so we find colleges that operate diversity-promoting admissions policies speciously accused of 'anti-White racism', with some facing financial penalties as a consequence (Quinn, 2025).

One British university, which had sought to outlaw 'transphobic propaganda' on its campus and require its staff to 'positively represent trans people' in course materials, was taken to task for its anti-hate initiatives, condemned for the modern-day thought-crime of 'stifling conservative viewpoints' (Jeffreys, 2025). As we will discuss in detail below, a formal regulatory inquiry would later respond to the backlash by fining the university for failing to facilitate 'free speech' (Adams, 2025a). But in culture-war contexts, the freedom (of other people) to say things is seldom defended to the death. Other colleges have been punished for permitting *too much* free speech on their campuses—or, at least, speech of the wrong kind—with the United States federal government punitively withholding funds from a number of institutions where students have chosen to exercise their freedom of speech to protest against right-wing establishment causes (e.g., Ahmadi, 2025).

Some political attacks on universities are more blatantly partisan than others. Brazil's Far-Right government dramatically slashed budgets of universities it claimed were beholden to 'left-wing influence', justifying the defunding with an almost self-satirising claim that 'ideological indoctrination' should never be supported by the state (Redden, 2019). (It is perhaps not entirely coincidental that it was in Brazil that, in the 1960s, Paulo Freire had promoted education as a tool for social justice.) In some instances, entire colleges have been absorbed into state control, ostensibly for teaching curricula deemed excessively 'woke' (e.g., Novak, 2021; Mazzei, 2023). And individual departments have also been targeted, often ones renowned for scholarship that examines, and thus threatens, geopolitical power hierarchies (e.g., Aratani, 2025).

Underlying the conservative critique of universities is a claim that academia, on the whole, has gone rogue. The complaint is that education has been displaced by indoctrination. Professors are blamed for contaminating young minds with piety and zeal. They are accused also of infantilising their students, of rendering them emotionally stunted and temperamentally deranged. The resultant moral panic can reach *Reds-under-the-bed* levels of paranoia. Colleges are portrayed as havens of 'cancel culture' (Ferguson, 2025) and 'political correctness' (Whittington, 2024),

Psychology's Quiet Conservatism

Brian M. Hughes

Psychology's Quiet Conservatism

How a Supposedly Woke Science
Promotes Capitalism
and Protects Privilege

Brian M. Hughes
School of Psychology
University of Galway
Galway, Ireland

ISBN 978-3-032-07723-3   ISBN 978-3-032-07724-0   (eBook)
https://doi.org/10.1007/978-3-032-07724-0

Cover illustration © eStudioCalamar

This Palgrave Macmillan imprint is published by the registered company Springer Nature Switzerland
AG.
The registered company address is: Gewerbestrasse 11, 6330 Cham, Switzerland

If disposing of this product, please recycle the paper.

Acknowledgements I am grateful to all at Palgrave Macmillan for their work on this book. I pay particular tribute to Beth Farrow for her foresight, dedication, and support throughout the journey. I also wish to acknowledge the input of the peer reviewers at various stages, which was informative and enhancing, and to thank Sophie Li for her assistance.

Many of the ideas on these pages were originally tested in classrooms and influenced by discussions with students and colleagues in psychology and beyond. For their ongoing inspiration and friendship, I wish to commend Siobhán Howard, Krys Kaniasty, Donncha O'Connell, Marguerite Hughes, and Ad Kaptein.

Finally, at the risk of self-plagiarism, I wish once more to thank my nearest and dearest for their ongoing patience and endurance as I toil through these never-ending endeavours. In this regard, specific shout-outs are due to Marguerite, Louis, and Annie.

Galway   Brian M. Hughes
October 2025

Also by Brian Hughes

Conceptual and Historical Issues in Psychology

Rethinking Psychology: Good Science,
Bad Science, Pseudoscience

Psychology in Crisis

The Psychology of Brexit: From Psychodrama
to Behavioural Science

A Conceptual History of Psychology:
The Mind Through Time

Contents

Part I   The Myth of Liberal Bias in Psychology   1

1 Universities Under Fire   3

2 Psychology at the Centre of Cancel Culture   11

3 The Claim of Liberal Bias in Psychology   23

4 The Problem With Claims About Psychology's 'Liberal Bias'   35

5 Culture War Psychology: Why the Liberal Bias Myth Persists, and Why It Is Damaging   43

6 Liberal Bias in Psychology: An Intellectual Mirage   55

Part II   Conservative Psychology, Past and Present   59

7 Legacy Conservatism   61

8 What the Standard History of Psychology Usually Ignores   67

9 Psychology's Roots in Theology   71

10 Psychology's Roots in Class Conflict   75

11 Psychology's Roots in Eugenics   81

12   Psychology's Conservative Paradigms   93

13   From Conservative Past to Neoliberal Present   109

Part III   Psychology, Capitalism, and Human Welfare   111

14   Hierarchies and Hysteria   113

15   The Contrivance of Capitalist Minds   119

16   Capitalist Psychology   125

17   The Capitalist Denial of Illness   133

18   'Personality Is Bad for You'   137

19   The Psychologising of the Sick   141

20   Unidentified Psychic Objects   151

21   Pathology and Protectionism   155

Part IV   Modernity and Declinism   163

22   Generation Snowflake   165

23   Depoliticising Youth Anxiety   173

24   Biological Reductionism Revisited   177

Part V   The Coddling of Conservative Minds   183

25   How Psychology Reinforces (and thus Perpetuates)
     Social Conservatism   185

26   Example #1: By Standing Up Against Safetyism   187

27   Example #2: By Pathologising Dissent   191

28   Example #3: By Labelling Deviance   195

29   Example #4: By Stigmatising Negativity   199

30   Example #5: By 'Othering' Ethnic Minorities   205

31   Example #6: By Policing Gender Identity   213

32   Example #7: By Perpetuating Traditional Gender
     Stereotypes   221

33   Example #8: By Exceptionalising Humanity   235

Part VI   Psychology's Whiteness Problem   245

34   Weird Science   247

35   Structural Racism in Psychology   253

36   Mechanisms of Whiteness   261

37   Silence as Supremacy   275

Part VII   Academic Exceptionalism and Psychology's
Blind Eye   279

38   Internalising the War on 'Woke'   281

39   Good Science, Bad Science, Pseudoscience   285

40   Exceptionalism in Psychology   291

Part VIII Beyond 'Liberal Bias': Four Paths
to a Well-Adjusted Psychology   297

41   Rights and Responsibilities   299

42   Path #1: Effortful Diversity   305

43   Path #2: Constructive Action in Education and
     Academia   309

44   Path #3: Constructive Action in the Public Square   315

45   Path #4: De-privileging Psychology   321

References   329

Part I

The Myth of Liberal Bias in Psychology

and as radicalisation camps for hard-left ideologies (Karpov, 2024). From the conservative perspective, modern universities are little more than adult daycare centres, places where future generations go to be brainwashed with extremist left-wing propaganda, and groomed to become hyper-triggerable, offence-seeking, virtue-signalling, emotionally fragile, climate-worshipping, gender-denying snowflakes.

Along Comes The Government

Several British political leaders have vociferously espoused such anti-university scepticism. In a 2019 speech to a think tank, a future home secretary derided academia for presenting Britain with 'a battle against cultural Marxism' (Kentish, 2019). The then conservative government duly undertook to 'crack down on universities' as part of the UK's own 'war on woke' (Ferguson, 2021). It announced a strategy to deal with the 'progressive monoculture' that had purportedly overtaken college campuses (Clarence-Smith, 2023), featuring plans to appoint a 'free speech champion' to the Office for Students, to whom complaints about campus liberalism could be forwarded. The prime minister later appointed a 'common sense tsar' to cabinet itself, with a brief to lead the government's 'anti-woke agenda' and to tackle 'the scourge of wokery' on university campuses (Addley, 2023).

Such sentiments are often reflected in broader political messaging. There have been speeches in parliament ridiculing the 'tofu-eating wokerati' (Davis, 2022), while a 'common sense group' of MPs have chided universities for wasting tax-payers' money on 'woke nonsense' and for 'poisoning the minds of generations to come' (Stringer, 2023). The election of an ostensibly left-wing UK government in 2024 did not entirely dampen campus concerns. The Overton window having shifted, the new administration quickly faced criticism over its choice to pursue hardline policies on a number of progressive issues, including the restriction of transgender healthcare (Perry, 2025), the cutting of disability benefits (Harris, 2024), and the planned deportation of immigrants (Griffiths, 2025). Their programme included a promise to retain the main elements of a law passed by the previous government that accelerated the clampdown on so-called university cancel culture (Adams, 2025b).

As it happened, 2024 was also the year in which the people of the United States re-elected as their president Donald J. Trump, perhaps the most stridently conservative politician ever to hold America's highest office. The Trump campaign had identified for itself many adversaries, but spent considerable energy berating the liberal tendencies of the country's university system. In so doing it espoused a wider conservative scepticism toward America's higher education institutions that had swelled over the preceding decade. By the time of the Trump re-election, several American states had already introduced laws aimed at snuffing out the supposed liberal takeover of education. Among the most active was the state of Florida, which famously attempted to legislate to prevent teachers from discussing sexual orientation and gender identity in class-rooms (Yurcaba, 2023). It also instituted a ban on the teaching of 'critical race theory' or 'related concepts such as intersectional-ity', on the basis that such ideas serve to 'push ideologies' on students (Thomason, 2023).

No fewer than eighteen US states have introduced legislation to limit education on 'divisive concepts', chiefly defined as any liberal perspective on race, gender, or sexual orientation. Penalties for breach-ing these laws include not only budget cuts for university adminis-trators, but also loss of tenure for individual faculty (American Association of University Professors, 2024). As a matter of legal statute, therefore, professors at American universities who attempt to explain structural racism, transexual identity, or Queer theory to their students could easily soon find themselves on the hunt for new employment.

Clearly, social conservatives believe that universities have been lost to the left. In their view, higher education has truly gone woke. The system, and therefore society, is in crisis.

Pining for Nostalgia

Attacks on university liberalism[1] are inherently sentimental. They rest on a complaint that the values being defiled are ancient and therefore pure. The critics do not wish to destroy our modern universities. Rather, they aim to *restore* these institutions to their former glory. They believe the universities to have been once distinguished, but lately desecrated. Their narrative amounts to an appeal to nostalgia.

For many centuries, universities were indeed safe havens for traditionalist views. Colleges would curate what was important about (Western) culture so that future generations could better cherish the universal norms of humanity. Scholars were expected to luxuriate in their thoughts, to ponder the esoteric and disregard the mundane. Erudition favoured theory, and often theology, over practicality, and new insights were circumscribed so as to avoid upsetting the religious and monarchical hierarchies around which society itself was organised. The mission of universities was to imprint knowledge, skills, and even virtue upon the young. These so-called ivory towers were, by reputation, quiet places, cut off from the noisy foibles of everyday living. They were academic enclaves where culture was to be upheld, not upended. It is to these halcyon eras that conservatives hark back. They wish to regain control of higher education. They want to make universities, and thus the world, great again.

None of this is to suggest that universities have never been seen as political. Quite the contrary. Universities, and the people who work in them, have always been recognised as instruments—indeed, as property—

---

[1] In this book, we will use the terms 'left' and 'liberal' as they are employed in the culture wars; that is, to label those who are liberal in the sense of being 'not socially conservative'. This is distinct from those who endorse the so-called Liberal movement within political philosophy, who tend to be more concerned with individual rights within the context of market economies, such as the right to own private property and to be free from government interference. This latter group also refer to themselves as 'liberals', but in a very different sense to the term that is used in culture war debates. In the context of the culture wars, the term 'liberal' is almost exclusively used to describe social liberalism, which focuses on social justice, societal obligations, equality, and inclusion, and espouses progressive stances on sociocultural issues, such as women's rights, reproductive rights, disability rights, and LGBTQ + rights. As such, it is this sense in which these terms will be used here.

of the government, and thus as part of the nation itself. It is from the endeavours of academic scholarship that populations learn who they collectively *are*. Universities are where a nation compiles its history and curates its art. It is universities that train bureaucrats, policymakers, legislators, and future leaders. Universities ensure a state's progress by stimulating scientific and technological innovation. And higher education prepares young people for employment and economic productivity. Universities are simply integral to the functioning of a country. They are, therefore, quintessentially political places.

Indeed, that is why universities are so often embroiled in political tension. The university is an infrastructure of the establishment, and academics are its operatives. To the wider (i.e., voting) public, the academic role is clearly mandated: academics are expected to assist new generations in carrying forward the standards of their forebears, not to actively foment the subversion of those very standards. They are duty-bound to enhance the student mind with facts, not to contaminate it with opinions. Academics are supposed to be *teachers*, not *preachers* (we have other people for that). So-called 'academic freedom' might sound like a fine idea to a professor, but to the outside world it is largely an internalist wonkish myth. Lecturers and professors are public *servants*; bound, not liberated, by their contract with society. They are expected therefore to live up to their side of the bargain.

The ivory towers are seldom quiet spaces nowadays. Today they have become battlefields, contested fronts in a noisy and global culture war. Faculty in fields such as politics, history, philosophy, and literature are regularly accused of bombarding their classrooms with leftist takes on the human experience, all frequently finding themselves in conservative crosshairs. This is especially true for those who teach psychology.

Throughout this book we will argue that although many (but by no means all) psychologists see themselves as liberal people, the *field* of psychology, as a body of knowledge, is far from 'liberally biased'. We will scrutinise claims of liberal bias in psychology and consider whether the available evidence supports such accusations, or whether in fact they are culture-war slurs propelled by partisan stereotyping. We will examine the deeply conservative roots from which modern psychology emerged, and how this history continues to influence the field today. And we will consider how modern psychology comprises a set of theories, methods, and practices that recurringly reinforce socially conservative worldviews. We will find that psychology is a geopolitically partisan discipline embedded within unjust societal structures and lumbered with heavy historical baggage.

As such, the sum of psychology is far greater than the individuals who partake in it. The personal views of individual psychologists do little other than distract from the political forces embedded within, and promoted by, their collective work. 'Liberal' psychology is an illusion. The progressive virtue that is commonly attributed to the field reflects signalling rather than substance.

Nonetheless, psychology's is the curriculum that truly spans the gamut of contemporary culture-war anxieties. Its faculty handle the hottest of hot button topics: from gender to genetics, from criminality to cognitive acuity, from brain development to bystander apathy, and from race to religiosity. It is the psychology syllabus that normalises diversity, that challenges stereotypes with data, and that sees souls as products of neurochemistry. On paper at least, psychology promotes human universality—the common interests and inherent equality of all humankind—as its core subject matter.

So, while the cut-and-thrust of the culture wars sees academia as a whole targeted with accusations of wokism, it is academic psychology that is regularly lambasted for being one of the wokest fields of all.

2

Psychology at the Centre of Cancel Culture

The notion of 'cancel culture' gained mainstream prominence in the late 2010s. The term is now familiarly used to describe a social climate in which a prominent person or organisation can suddenly be ostracised if they behave in a manner deemed unacceptable by a wider group. But cancel culture is far from a new idea. Transgressors have long been boycotted, shunned, or ostracised by their communities. Nonviolent boycotts targeting corporations date back at least to the 1700s, when they were used to publicly shame businesses that profited from the slave trade (Holcomb, 2016). Indeed, the term 'boycott' itself was coined in honour of a nineteenth-century English land agent, one Charles C. Boycott, after inhabitants of an Irish village conducted an organised campaign to isolate him in protest at what were seen as unfair local evictions (Moran, 1985).

In a more literal sense, the language of being 'cancelled' originated in the 1980s among speakers of African-American Vernacular English, before circulating more broadly through Hip Hop fandom. It was conspicuously used in the cult crime movie *New Jack City* (Van Peebles, 1991), in which the lead character, played by Wesley Snipes, 'cancels' his girlfriend after she criticises his criminal lifestyle

(Rogers, 1993). Having become popular in Black American culture, the term spread widely as a way to describe the calling out of racism in the United States, especially among Black American social media users (Ng, 2022). It was within the milieu of social media hashtags that the idea of 'cancelling' those who stepped out of line gained a wider audience. The term became central to campaigns such as #MeToo and #BlackLivesMatter.

As such, cancel culture is more than just speech suppression or deplatforming. It is a particular form of political protest that is rooted in minority activism. In this sense, choosing to condemn cancel culture is an inherently conservative reflex. It reflects the efforts of those with privilege to push back on dissenters. Cancel culture is the David to privilege's Goliath. After all, speakers tend to have much more social capital than their audiences: when audiences rise up to exert the only power they have—that of refusing to provide speakers with attention—it represents a form of protest multiplication. The anonymous masses are empowered in combining their leverage. They become capable of toppling those who might oppress them.

The idea that cancelling is now so prevalent as to represent a 'culture' in its own right is a recurring conservative trope. It arises from a dual accusation, both elements of which are distinctly tenuous: firstly, that cancel culture has run so rampant as to have laid siege to the public square; and secondly, that cancelling is performed almost entirely by left-wing liberals (Vivian, 2023). The subtext is that cancel culture, or cancelling more generally, is simply unfair on conservatives: it is alleged to involve the zealous targeting of undeserving victims in the service of those with ulterior partisan motives.

Much of this supposedly out-of-control cancelling is said to be aimed at political or cultural opinions that are unpalatable to the political left. Cancel culture is blamed for making scapegoats of a wide range of public figures, from celebrities to politicians to CEOs. Conservative academics, too, it is claimed, are typically targeted, be they scientists, economists, historians, or literature professors. In fact, it is arguably the college-campus iteration of cancel culture that is complained about the most. Academic venues regularly serve as proxy sites for partisan political conflicts (Vivian, 2023).

Topic boundaries are frequently fuzzy, and academics who find themselves cancelled will often claim to be left-leaning in their overall political worldviews. But the controversies that lead to calls for their cancellation are seldom that varied. While many issues can arise, some seem to arise much more than others. The ones that cause the greatest strife, either directly or indirectly, tend to centre on the subject matter of psychology. This is what puts psychology at the heart of the culture wars.

The First of Two Examples: Gender Identity Scepticism as a Cause Célèbre

One of the most incendiary of these hot button issues relates to gender identity. More specifically, it relates to the question of how human beings subjectively and personally experience their own 'gender', and whether that experience is (or should be) related to how *other people* view their 'sex'. The concept is frequently discussed with a specific focus on transgender people and their rights. Despite the voluminous media coverage dedicated to their existence, transgender people make up a minuscule fraction of the population. While survey methods vary, most estimates suggest that around 0.3% of the human population will self-identify as transgender. The number who receive gender affirming surgery, hormonal therapy, or other forms of transgender-related medical care is even lower: in countries where these options are available, only around 0.009% of the population—or 9 of every 100,000 people—will ever seek them out (Collin et al., 2016).

Nonetheless, gender identity is frequently a trigger for campus conflict. A key case concerned Kathleen Stock, a former philosophy professor at Sussex University. While at Sussex, Stock became embroiled in controversy over her links to the LGB Alliance, a British campaign group that had frequently been accused of discriminating against trans people. She had also signed a public declaration written by the Women's Human Rights Campaign, an organisation that had previously called for 'the elimination' of 'the practice of transgenderism' (Parsons, 2021). The LGB Alliance and the Women's Human Rights Campaign have both been described as 'hate groups' and criticised for their extremist positions (Hinsliff, 2021; Wakefield, 2022; Southern Poverty Law Center, 2023). As such, student representatives claimed that Stock's support for these initiatives exceeded the remit of her academic freedom. They felt it violated the university's staff code by 'demonstrating hostility towards … individuals or groups of individuals by reason of a protected characteristic (as defined in the Equality Act 2010)' (Lavery, 2021).

While Stock herself maintained that 'to say males cannot be women … is certainly not hate speech,' that view continues to be disputed. For example, Alexander Brown and Adriana Sinclair, hate speech experts who teach at the University of East Anglia, have pointed out that such statements are structured as 'identity attacks' and so very often do meet the ordinary definition of hate speech (Brown & Sinclair, 2024). In other words, Stock herself—a philosophy professor—might believe the law on speech to be different, but for legal scholars, it was entirely reasonable for her students to raise hate speech concerns. The fact that the British supreme court has since ruled that the term 'woman', when used within the Equality Act 2010, refers to 'biological and not certificated sex' (Malik, 2025) does not legitimise the slandering of transgender people with hurtful commentary.

Stock framed her views as those of a philosopher. But several of her philosophy peers found fault with her analysis. One of their criticisms related to her failure to account for excluded cases. For example, while Stock had many concerns about trans women sharing female-only spaces with cisgender women, the remedies she offered failed to account for the existence of trans *men*. Stock objected to 'biological' males being admitted to female-only spaces. But by her own reasoning, *trans* men should be admitted to those spaces on the basis of being 'biologically' female—even if they possess male genitalia by virtue of having transitioned to male gender. By Stock's own logic, therefore, some people with beards and penises *could* use female bathrooms—and indeed should be *required* to do so (Zanghellini, 2020). Stock was also criticised for ignoring the extensive work on gender classification that other philosophers had produced over many decades. As a result, many of her more strident conclusions failed to take account of complexities that had long been highlighted by scholars who actually specialised in this area of philosophy. In other words, Stock's arguments on gender classification were rudimentary, and her reflections on gender identity came across as naive. One philosopher compared her tone to that of a student writing an essay about the definition of a table (Bettchner, 2018).

Alongside these scholarly critiques, the students maintained their campaign of protest against Stock's links to the LGB Alliance and Women's Human Rights Campaign. The leadership at Sussex vehemently defended Stock and refused to fire her. But Stock eventually concluded that the constant criticism was no longer tolerable. When her own trade union called on her university to take a 'strong stance against transphobia,' she felt she would be unable to continue to work at the institution (Adams, 2021). She instead decided to resign. It was Sussex University that would later be fined for failing to defend 'free speech'.

Given the news coverage of her case, Stock was quickly hailed by the right-wing media as a celebrity victim, a newly beatified cancel culture martyr. Among other roles, she was offered a job as founding faculty fellow at the University of Austin in Texas, a private liberal arts college established in 2021 to much public fanfare as an 'anti-woke' alternative to mainstream higher education (Place, 2021). She also received an OBE, one of the most prestigious awards in the UK's civil honours system, for 'services to higher education' (Le Duc, 2022). She continues to maintain a significant profile in the media and public life, campaigning, writing, and speaking widely on gender issues.

Critics of campus cancel culture have long suggested that Stock was physically endangered by the student protests, and that she needed to leave her job in order to ensure her own safety (Griffiths, 2021). An irony here is that while very few trans-sceptic (or so-called 'gender critical') commentators have ever been physically harmed, transgender people are extremely likely to experience significant violence. For example, statistics from the US Federal Bureau of Investigation show that transgender people are victimised in 3.2% of America's hate crimes, despite comprising just 0.3% of the country's population (Federal Bureau of Investigation, 2023). Transgender people are by far the most disproportionately targeted minority group included in the FBI's hate crime data. Every two weeks, a trans person in the US is murdered (Human Rights Campaign Foundation, 2023). Far from being a threat to others, it is arguably trans people who find themselves truly endangered, a predicament only exacerbated by the whipping up of anti-trans sentiment in the communities in which they live.

It would be an understatement to say that transgender people are frequently problematised. Transgender individuals regularly hear their very existence discussed as though it represents an immediate threat to society, one requiring an urgent social policy response. The way gender is experienced by individuals, how this might relate to physiological processes, and the role of feedback and social perceptions in impacting upon a person's well-being, brings the question of gender identity very much into the realm of psychology. How psychologists discuss gender actually matters (or, at least, *should* matter).

In Part V, we will discuss psychology's role in policing gender identity. We will examine the norms psychologists use and how they have come to use them. But for now it is of interest to note that while gender identity is about as 'core' as core psychology subject matter can be, many of the most vocal commentators on this topic are non-psychologists. While everyone is entitled to have their say, it does mean that public discourse usually takes little or no account of what research in psychology has uncovered about the nature of gender identity, or about the lives and experiences of transgender people. Instead, discourse revolves around psychologically decontextualized debates concerning what is often simplistically labelled as 'the trans issue'. Ultimately, whether psychologists have been effective at contributing any useful, empirically-grounded knowledge to these debates is very questionable.

A Second Example: The Trope of Human Hierarchy

A second category of campus cancel-culture controversy relates to the issue of human hierarchies. This is the term used to describe claims that some human groups are inherently more capable than others (and, by implication, that some groups are naturally *less* capable). One example of an appeal to human hierarchy is the claim that men have greater leadership abilities than women, because it suggests that around half the human population is, on average, better at leadership than the other half. Likewise, claims that Ashkenazi Jews enjoy 'an advantage in average intelligence' (Pinker, 2006) compared to other people is also an appeal to human hierarchy, because it implies that some members of the population will be cognitively more capable simply by virtue of belonging to a particular ethnic group.

A key problem with claims about human hierarchies is that they are frequently groundless. For example, the vast majority of so-called sex differences in ability have proven impossible to corroborate with empirical evidence. This includes virtually all popular cultural stereotypes about the relative skills of women and men (Hyde, 2005, 2014; Hyde et al., 2019). It appears that gender stereotypes persist because of the way rigid social prejudices are culturally reinforced. Likewise, claims of race differences in psychological attributes have been discredited for many decades. The continuing assertion that race differences have been proven by 'scientific' research belies the fact that the research being referred to lacks any meaningful validity, reliability, or rigour. Simply referring to research as 'scientific' does not make it so. In fact, it is quite possible for a study to convey the *appearance* of science without being scientific at all. This is why the study of race differences in psychology is widely considered to meet the definition of pseudoscience (Hughes, 2023; Rutherford, 2022).

Therefore, it should be no surprise that anyone who makes unsubstantiable claims about human hierarchies can expect to face criticism. Apart from dallying with pseudoscience (or, at best, turning a blind eye to bad science), casually hypothesising about sex or race differences often just comes across as sexist or racist. Scientists or scholars who attempt to leverage their general academic standing in order to put forward highly specialised claims about these matters risk a further accusation: that they are failing to make clear they are speaking *outside* their particular domain of expertise. Exceeding one's academic competence is more than just a scholarly faux pas. It is a breach of professional ethics. This is because it will likely mislead audiences into thinking that contentious views are somehow supported by rigorous professional knowledge, when in fact they are little more than hot takes, provocative ideas that reflect the speaker's personal beliefs, values, and biases.

Even Nobel laureates and renowned scientific pioneers can be guilty of this type of infraction. The American biologist James Watson is a notable case in point. Along with the English scientist Francis Crick, Watson was credited with establishing the double helix structure of DNA in the 1950s. The duo's research revolutionised the way science understands the nature of living material, an achievement that led to the award of the 1962 Nobel Prize for Physiology and Medicine. However, many decades later, Watson began to speak about issues that were quite detached from the microscopic subject matter of molecular biology.

Despite having no training in the social sciences and never having conducted research on behavioural psychology, Watson saw fit to put forward a number of claims relating to supposed race differences in psychological attributes. He declared that there were significant race differences in intelligence, temperament, sex drive, creativity, submissiveness, and other traits. He specifically suggested, without direct evidence, that these race differences were grounded in underlying genetic processes. And he dramatically theorised that government policies that failed to take account of race differences would eventually destroy society. In short, Watson proposed an extensive meta-theory that spanned multiple areas of specialist expertise, encompassing aspects of biological psychology, social psychology, educational psychology, psychometrics, sociology, political science, anthropology, and public administration. The problem, of course, was that his knowledge of these fields was lacking. Watson cited studies to support his assertions, but relied on data that are well known to be questionable and which do not substantiate the claims advanced by race-difference advocates (Panofsky et al., 2021). Researchers who specialise in these areas understand that claims such as Watson's are specious.

Watson denies being a racist. However, he appears committed to his view that society is being destroyed by the failure to acknowledge race differences in psychological variables, despite not being able to sub-stantiate that view with meaningful data. As a result of his reputation, Watson has found himself marginalised in the academic community. Few institutions are interested in having him speak on their campuses. If anything, his public statements have become more controversial over time. He was suspended from his role at the prestigious Cold Spring Harbor Laboratory in New York, and eventually stripped of all honorary titles the laboratory had given him (Durkin, 2019). Watson, to use the phrase, has been well and truly cancelled.

In this regard he is not alone. Academics around the world have faced criticism for espousing similar views. In 2019, Noah Carl, a sociologist who worked at Cambridge University, had his research fellowship terminated after he made several public statements regarding supposed race differences in intelligence and criminality. A formal university investigation found that his research exhibited crucial methodological flaws and represented 'poor scholarship'. Moreover, it revealed that Carl had collaborated with 'a number of individuals who were known to hold extremist views,' which placed his work 'outside any protection that might otherwise be claimed for academic freedom of speech' (Bradbury & Cook, 2019). Carl briefly pursued a legal challenge to his dismissal, aided by an organisation that specialised in providing legal services to White nationalists and neo-Nazis in the United States (Bradbury & van der Merwe, 2019). His challenge was eventually withdrawn. Nowadays, critics of university cancel culture frequently report the episode as an example of overzealous free-speech suppression by universities, but they typically omit to mention Carl's links with off-campus extremists (e.g., Stevens et al., 2020). For example, an editorial in *The Times* complained that Carl's only crime was to 'challenge the "woke" left-wing orthodoxy' (*The Times*, 2019).

The following year, a physicist at Michigan State University, Stephen Hsu, was similarly forced out of his role as university vice president, after making several contentious claims about race differences. Once again, critics of cancel culture frequently report Hsu's case very selectively, ignoring the fact that Hsu had promoted his ideas on a well-known White supremacist podcast (Lyons, 2020). A writer in the *Wall Street Journal* ignored his involvement with White supremacists, and complained that Hsu was simply 'taken down' by an irrationally offended 'Twitter mob' (Melchior, 2020). Focusing on the removal of a person from their job, while glossing over the reasons for that removal, reinforces the stereotype of universities as politically intolerant places. It suggests they are partisan institutions that punish dissenters merely 'for expressing ideas' (e.g., Jussim, 2019).

Similar controversies have arisen with regard to claims about sex differences. James Watson himself attracted much criticism for his views on the abilities of women, including his belief that women are less numerate than men and therefore make 'less effective' scientists (Gabbatiss, 2019). While Watson's comments probably reflected little more than off-the-cuff misogyny, other academics have sought to present an overarching argument based on statistical reasoning. A common approach has been to cite what has become known as the 'variability hypothesis', the idea that men vary in their abilities much more than women do.

The variability hypothesis suggests that while some men are highly proficient, other men are extremely incompetent, whereas women generally cluster around the middle of the range. At first glance, the theory seems to acknowledge that many men will be inferior to most women. But few competitions are decided on the basis of who is the most incompetent. The crucial part of the hypothesis is what it proposes about the *high* end of the ability spectrum. With men's abilities said to be more variable, the theory proposes that in any competitive process—for example, job interviews, academic tests, or entrepreneurial markets—we should always encounter more successful men than successful women. By implication, therefore, any competitive process that produces (or that is designed to produce) *equal* success for women and men must be somehow flawed.

The variability hypothesis frames an interesting story about the possibility of human heterogeneity. However, it is far from clear that it is actually true. A key problem is that it is almost impossible to test with real data. Almost any dataset describing job interviews, educational tests, or similar competitive processes will contain only that subset of women and men who, for cultural reasons, have been presented with the *opportunity* to participate. It is self-evidently the case that not all women and men will enjoy such opportunities.

Any sex difference in outcomes could be caused by an initial divergence. If more men than women gain access to engineering degree programmes, it does not mean that men are 'better' than women at engineering. It could just mean that men are more likely to be culturally reinforced to *choose* engineering as a career path. The divergence will widen over time if, for example, more women avoid engineering *precisely because* they fear that the preponderance of male engineers will present them with a hostile work environment. In short, it will always be logically impossible to attribute disparate outcomes directly to people's sex, until such time that we can be sure forces such as cultural sexism or sex biases no longer exist.

Some researchers persist in claiming that there is statistical evidence for the variability hypothesis. They point to large datasets where they feel cultural biases are not in play, such as results from school-administered tests that include every single child in a school system, rather than a selected sample of children. However, even these datasets are not immune from cultural biases. Whenever multiple such datasets are compared, it is usually revealed that so-called male variability turns out to be associated with contextual features, such as school type, national income, societal gender discrimination, and even religion (Kane & Mertz, 2012). As such, not only is it *logically impossible* to exclude these considerations, it is *empirically demonstrable* that cultural contextual factors are liable to create the *appearance* of male variability where such variability does not truly exist.

In January 2005, Lawrence Summers, an economist who at the time was serving as president of Harvard University, spoke at a conference on sex differences in professional science and engineering. In his speech, he used the variability hypothesis to explain why he believed elite universities employed fewer women than men. What was especially controversial about his speech was that Summers went further than simply citing a questionable hypothesis—he explicitly rejected the possibility that cultural factors might be involved in women's career prospects at all. He dismissed out of hand the idea that women's success might be impeded either by discrimination or by differences in socialisation. He rejected the view that some women might rationally choose to avoid potentially hostile male-dominated workplaces. And he decried the argument that women could ever be disadvantaged by inadequate allowances for child-care or maternity leave.

According to Summers, sex differences in professional science and engineering were entirely unrelated to sexism in society. Instead, he argued that the underrepresentation of women in elite universities such as Harvard was driven by differences in 'intrinsic aptitude, and particularly of the variability of aptitude.' He suggested that he was fully in favour of gender equality and 'would like nothing better than to be proved wrong' (Rimer & Healy, 2005)—but as far as he was concerned, the problem of female under-representation could not be resolved by wishful thinking.

His remarks were immediately and widely criticised, not only for their sexist tone but also for their scholarly carelessness (Hemel, 2005). Summers did not address, or even mention, the shortcomings of the variability hypothesis. Essentially, he framed it as a fact. Nor did he present any empirical evidence to support his questionable claim that cultural factors (such as sex discrimination) played no role. Instead he ignored—or was perhaps oblivious to—the extensive research evidence that shows how social biases impact upon the career progression of women. Critics quickly pointed out that since Summers took over as president, the proportion of women offered tenured faculty positions at Harvard had fallen from 36 to 12.5%, with just four women (but 28 men) being hired in the year preceding his speech (Goldenberg, 2005).

Summers rode out the immediate backlash, but within a year he had resigned from the presidency. His case is frequently identified as the archetypal anti-conservative academic ousting, an early example of what would become known on campus as 'wokeism'.

3

The Claim of Liberal Bias in Psychology

Few academic fields stand accused of cultural thought crimes more than psychology. While philosophers, historians, economists, or sociologists might choose to *discuss* such topics as gender identity, race, intelligence, or sex differences, it is psychologists whose discipline is premised on the *empirical explanation* of these things. It is psychology that generates the first-principles evidence base that tells us *how* gender identity operates, *what* intelligence is, and *why* behaviour might differ by sex.

Of course, the knowledge produced by psychologists will always be tentative. The subject matter of psychology is not the same as that of physics, where critical materials can be weighed in nanograms or photographed while under a microscope. Psychology requires professional judgement to determine what the variables even *are*, never mind how they can best be measured. Psychologists are not the only scientists who deal with abstractions, but they do tend to deal with more abstractions than most.

All sciences rely on human interpretation of data. But not all sciences deal with data that are entwined with controversial social issues of the day or with contentious public policy debates. Gender identity and human hierarchies are not the only psychological fronts in today's culture wars. In fact, just about all culture-war causes are affected by what psychologists have to say about the nature of the underlying issues.

Debates about abortion, for example, revolve around contentious claims and counter-claims regarding the mental health impacts of terminating a pregnancy, the risks posed by stress and suicidality, the autonomy of choice-making, the cognitive maturity of pregnant teenagers, the coercive influence of politically biased physicians, the effects of social stigmatisation, and even the nature of human consciousness and whether it can be said to exist in an embryo. All these considerations invoke subject matter informed by research and scholarship in psychology. In many forums, the process occurs literally. Political debates about reproductive rights frequently reduce to bitter tit-for-tat exchanges in which alternating barrages of psychology research studies are flung, as ammunition, by each side against the other.

The same can be said for other culture war battlefields. Psychology concepts are deployed as weapons—by non-psychologists, in the main—in fights over teenage sexuality, contraception availability, fertility treatment, sex education, vaccination programmes, surrogacy, microaggressions, trigger warnings, privilege, media depictions of women, the rehabilitation of criminals, the permissibility of xenophobia, the merits of traditionalism, the trustworthiness of politicians, the lived experiences of minorities, remedies for the climate emergency, whether there really is a climate emergency or whether we are just being 'gaslit' about it, whether a government should tell its people what to do during a global pandemic, or even *how* a government should best determine those rules and whether people should be obliged to obey them.

Psychology is often referred to as the science of behaviour. Culture wars are fought over how people actually—or *should*—behave.

Psychology and the 'Academic Left'

Over the past two decades, conservative critics—both within the field of psychology and outside it—have bemoaned the manner in which psychologists have ostensibly been overwhelmed by intellectual activism and unreasoned social consciousness. On every culture war divide, it is said, psychologists use their scholarly discretion to *arbitrarily* frame their knowledge base. They draw overstated conclusions from ambiguous data, in order to trump up 'evidence' for what are partisan (and reckless) left-leaning positions.

Mainstream media and academic journals alike present a familiar set of stereotypes: the psychology curriculum as liberal dogma, and professional psychologists as woke social justice warriors and liberal 'snowflakes'. These left-leaning psychology academics are supposedly not satisfied with merely adulterating the curriculum that they impart to their students. They are also intent on driving out anyone with competing views. Theirs is a mission of mind control, it is said, in which threats to political purity are legitimate targets for opprobrium.

The claim that academic psychology is 'liberally biased' has had a slow-burning but persistent modern history. It is intertwined with a wider critique of university education as a whole. Some of this criticism was amplified by the so-called 'science wars' of the early 1990s, a campus controversy that arose after prominent American scientists expressed alarm about what their colleagues in the humanities were teaching (e.g., Gross & Levitt, 1994). They complained that in fields such as literary criticism and philosophy, their fellow academics were polluting classrooms with scientific denialism. These renegade professors, the scientists argued, were seeding scientific illiteracy across society at large.

The scientists took particular umbrage at 'postmodernism', a scholarly approach that encourages students to question the implied certainty of knowledge. Postmodernism is intended to neutralise the naivety of *modernism*, the unwavering belief that our modern world is guaranteed to benefit from technological advances and scientific certainty. The postmodernist view questions this faith in the robustness of knowledge. It urges caution rather than confidence. It emphasises the core (psychological) fact that all information passed from one human to another, including every-thing in science, is subject to interpretation—it makes sense only in terms of how those humans *perceive* the world. Postmodernism points out that people sometimes, if not often, come to mistakenly believe something to be true when it is, in fact, false.

The science war critics, who often referred to themselves as 'scientific realists', dismissed all these points. They ridiculed their fellow professors for arguing that much, if not all, human knowledge might in fact be a delusion. For them, this was an intellectual extravagance. It made no sense, they felt, to reject objectivity, or to claim to doubt *all* ideas that were previously considered to be factual. Gravity, astronomy, aeronau-tics, physiology, microbiology, immunology, chemistry, meteorology—these were fact-based disciplines, the scientists argued, not social con-structions based on opinions-treated-as-facts. The scientists complained that postmodernism was nothing less than an 'abuse of science', an intellectual fantasy that qualified as 'fashionable nonsense' (Sokal & Bricmont, 1998).

While the science wars are now often framed as a critique of the humanities, the targets of these skirmishes were actually quite wide-ranging. Disciplines far beyond the philosophy and literature depart-ments found their scholarship suddenly skewered. Critics aimed fire at any area that emphasised the role of perception and interpretation in the production of knowledge. Culprits included cultural studies and women's studies, but also environmental sociology and public health (Gross & Levitt, 1994). The attacks encompassed three primary argu-ments: firstly, that mainstream science was an unstoppable intellectual force that had improved the world with empiricism; secondly, that non-scientists were naively ignorant of—and therefore poorly qualified to criticise—the highly complex methods used by professional scientists; and thirdly, that non-scientist professors who deigned to question the validity of what was published in scientific journals had no basis, other than whimsy and arrogance, on which to do so.

A key problem with this critique is that it overestimated the rigour of scientific knowledge production. We are now much more aware of the flaws that plague professional science, and of how its outputs can be hampered by measurement error, commercial entanglements, culturally reinforced assumptions, and a variety of ordinary human frailties (we will return to some of them in Part VII). Not all criticisms of science are meritorious, but the science warriors' idea that science should occupy an academic pedestal—*beyond* the criticism of non-scientists—seems, in retrospect, to have been somewhat wrong-headed.

An important legacy of the science wars was the terminology it bequeathed. The language used by the scientific realists to label their on-campus opponents framed matters in a targeted way. As biologist Paul R. Gross and mathematician Norman Levitt put it in the subtitle of *Higher Superstition*, their best-selling book on the science wars, the quarrelsome sceptics who questioned scientific truths should be regarded as the 'academic left' (Gross & Levitt, 1994). Scholars and researchers who dared to question the certainties of past knowledge—to ask whether 'facts' might sometimes be biased as a result of the non-neutral contexts in which they are first written down—were no longer seen as merely academic rivals. They were cast as *political* adversaries.

The Liberal Bias Claim

Gross and Levitt explicitly identified psychology as one of the subject areas subsumed by the academic left, listing it as one of the fields that 'have had to come to terms with a militant, sometimes angry challenge to their settled ways of doing business' (p. 107). But, in reality, the complaint that psychology is a leftist profession long predated the science wars. Psychology's 'settled way of doing business' had attracted criticism from conservatives for decades prior to this supposedly emergent challenge.

Since at least as far back as the 1960s, critics have cited US survey data to claim that psychology, as a field, is disproportionately left-leaning. One poll published in 1965 was pivotal in framing the debate. Psychologists working at American universities were asked about their electoral voting preferences. By far the majority said they voted for the Democratic Party, whereas only a minority reported that they voted Republican (McClintock et al., 1965). This was seen as evidence of a problematic skew within psychology. As the Democratic party was considered to represent the political 'left' and to espouse so-called 'liberal' or 'progressive' social values, voting for Democrats in such large numbers was taken to mean that psychologists were biased toward liberalism.

Despite being sixty years old, this historical finding continues to be mentioned today by those who wish to marshal evidence of psychology's supposed political skew (e.g., Redding, 2023). But the exact meaning of such voting patterns can be difficult to discern. For one thing, it is important to take account of the electoral base-rate of the time. In 1965, the Democrats were by far the most popular political movement in the US, having just won their second presidential election in a row, romping home with the support of almost two-thirds of voters. In other words, the voting preferences of psychologists simply resembled those of American electors more generally. It did not evince any particular 'bias' per se. In fact, if psychologists had been split evenly between Democrats and Republicans, then that *would* have been evidence of a literal skew within the field. It would have meant that psychologists, as a group, held views that drew more from minority than majority opinion. As it was, psychologists were just the same as everyone else.

In any event, voting patterns do not always reveal authentic political preferences. It is important to note that the United States operates what is essentially a forced-choice binary electoral system. Voters are not invited to express their philosophies in the polling booth, but are required to select from the two main options available, even if they dislike both. To much of the world outside the US, both Republicans and Democrats represent alternative strains of conservatism (Hibig & Moshagen, 2015). This was as true six decades ago as it is today. While the 1960s Republican party were indeed firmly right-wing, Democratic presidential candidates of that period were also markedly conservative. They ran campaigns largely centred on promises to increase military spending and adopted hostile stances against socialist economics, stances that today would be seen as emblematic of right-wing politics. Their liberalism, such as it was, was pretty low-key.

The same was true of professional psychologists. Psychologists in the 1960s rarely advocated for liberal positions on social issues. Indeed, when it came to the Civil Rights movement, one of the major progressive causes of the time, the American Psychological Association actively *discouraged* its members from publicly supporting racial equality (Williams, 1974; we will revisit these events in Part VI). Overall, when viewed in context, the political views of psychologists in the 1960s offer very weak evidence for claims that psychology has been dominated—for generations, no less—by a stridently liberal ethos.

Making the Modern Case

Modern claims of liberal bias in psychology are barely more sophisticated than efforts to draw inferences from sixty-year-old voting patterns. Take, for example, one recently published book that ostensibly undertakes to expand upon the problem—*Ideological and Political Bias in Psychology: Nature, Scope, and Solutions*, edited by Frisby et al. (2023). Spanning more than 950 pages, its thirty-one chapters feature contributions from a variety of authors describing sundry instances of purported political bias, ranging from how some psychologists discuss authoritarianism (Costello, 2023), to how other psychologists advise parents on whether to punish their children (Larzelere et al., 2023), to the way others use neutral phrases when describing gender (Bailey, 2023).

However, despite the ample page count, neither editors nor contributors present significant evidence to support the book's central claims: that the examples presented are indicative of psychology as a whole, or that the field of psychology, in general, is politically biased. A selection of polemical essays and case studies (even one verging on a thousand pages) is of little probative value on its own. As an edited volume of invited contributions, the content of such a book is *by definition* going to be selective. One might even suspect it is biased.

The one chapter to offer data directly pertaining to the prevalence of liberal bias in psychology was authored by Richard E. Redding of Chapman University in California (Redding, 2023). Redding, who also serves as one of the volume's four editors, has written many times before about liberalism in psychology. His career-launching contribution was a widely read *American Psychologist* article, published in 2001, that called for greater 'sociopolitical diversity' in psychology, given the field's 'conservative absence' (Redding, 2001). The article warned of the consequences of this conservative absence, arguing that standards in psychology will inexorably decline unless conservatives are invited in from the margins. As a pioneer in the field, Redding was presumably excellently placed to provide an up-to-date account of the evidence showing just how prevalent psychology's liberal bias actually is.

So, What is the Evidence of Liberal Bias in Psychology?

Redding's chapter begins with a notably pre-emptive comment. 'It is well understood among psychologists,' he writes, 'that a liberal political ethos pervades the discipline' (p. 79). In this form, Redding presents his conclusion as his premise. The chapter poses not a literal question, but a rhetorical one, to which the answer is predetermined. With such an agenda having been set, the ensuing text need only showcase those findings Redding feels support his point.

The exercise unfolds over four main sections. The first section focuses on studies that examine the political attitudes of professors, and we will scrutinise this section in detail below. The other sections, however, deal with evidence that seems peripheral at best, and, at worst, irrelevant.

For example, the second and third sections focus on studies of psychology students and psychology practitioners, including therapists. It is somewhat questionable whether students and therapists are truly relevant to the overarching claim that psychology is afflicted by a liberal bias. Firstly, it is unclear how the political attitudes of students might be used to divine whether *the subject matter of the curriculum that is presented to them* is itself politically biased. Students might well hold strong political views, but classes are delivered by teachers, not by students, and university classrooms often contain students who dislike the subject matter that is being taught. Secondly, for similar reasons, the political attitudes of psychotherapists, many of whom last attended college several years ago, seems a poor basis on which to judge whether the field of psychology is skewed by supposedly woke scholarship.

The fourth section in Redding's chapter is even more removed from its intended target. It considers the political attitudes not of psychology academics, students, or indeed therapists, but of 'consumers of psychological services' (p. 84). These consumers are said primarily to comprise psychotherapy patients. Redding argues that if it can be shown that only liberals seek out therapy, then this would expose how psychology's value system is unappealing to conservatives. From this we might then conclude that the field of psychology is politically biased. But again, it is unclear whether this inference is sound. It could well be that conservatives avoid psychotherapy, especially if they *perceive* it to be biased. But unless they attend therapy, how could they possibly know for sure? Their perceptions might just be wrong. (In such a scenario, it would be more persuasive to argue that it is *conservatives* who are biased against *psychology*, rather than the other way around.) As it happens, Redding concludes by pointing out that there has been 'virtually no research on the political attitudes of those who consume psychological services' (p. 84)—so for all we know psychotherapy clinics might, in fact, be *overflowing* with conservative clients. In any event, the merits of conducting such research, for now, remain hypothetical.

This leaves just the first section of Redding's chapter—the section dealing with the politics of psychology professors—to fulfil his aim of showcasing evidence of the field's liberal bias. Unlike students, therapists, or patients, it is professors (or academics) who shape the field. They do this by conducting research, by formulating findings, and by designing and delivering curricula. If psychology is liberally biased—if it is non-neutral in how it deals with (say) gender, racism, or sex differences—then this should be revealed by examining the attitudes of those who actually *produce* the field's knowledge base, rather than by soliciting the views of students, practitioners, or others who end up consuming that knowledge, without actually being responsible for it.

Redding's section on psychology professors includes two sets of studies. The first comprises a series of opinion polls conducted on members of different psychology organisations: specifically, three surveys of the Society for Personality and Social Psychology (SPSP) and one of the Society of Experimental Social Psychology (SESP). Both societies are ostensibly international, although the majority of their members are located in the United States.

The first two SPSP surveys were conducted in the early 2010s by Yoel Inbar and Joris Lammers, social psychologists who at the time were working at Tilburg University in the Netherlands. Their results were published as part of a special feature in the journal *Perspectives on Psychological Science* (Inbar & Lammers, 2012). The paper has since become the most frequently cited work, by far, on the question of liberal bias in psychology. Its findings made for explosive reading. Inbar and Lammers reported that some 90% of SPSP members self-identified as 'liberal' on social issues, whereas only four per cent identified as 'conservative'. More controversially, they found that nearly 40% of SPSP members admitted that they were willing to exercise political discrimination in workplace contexts. Specifically, four-in-ten respondents were happy to declare that they would deliberately refuse to hire a job applicant if that person was a conservative.

The third SPSP survey was conducted by the SPSP itself (Society for Personality and Social Psychology, 2019). It revealed a similar pattern. Redding states that 89.6% of SPSP members in this survey described themselves as 'liberal', with just 4.4% claiming to be 'conservative'. The final survey covered members of the SESP (Buss & von Hippel, 2018). According to Redding, this survey found that 89% of SESP members consider themselves 'left-of-center' politically, whereas only 2.5% describe themselves as 'right-of-center'.

Alongside these membership surveys, Redding outlines a second set of analyses produced by the business school academic Mitchell Langbert. Langbert has achieved fame for collating the registered political affiliations of faculty members working in US colleges. According to Redding, in half of the country's top forty universities, Langbert was unable to find any registered Republican working as a faculty member in psychology (Langbert et al., 2016). He also asserts that when Langbert examined the four highest ranking colleges, the ratio of Democrats to Republicans in their psychology departments was 11.5 to 1 (Langbert & Stevens, 2020).

While Redding goes on to describe other studies looking at the 'professoriate generally'—that is, academics as a whole rather than academics who teach psychology—he presents no further data on the political orientations of psychology professors. It would appear that, as far as Redding can argue, the above-mentioned handful of opinion surveys and voter audits comprises the entirety (or at least the best) of the evidence. While anecdotes might well abound, it seems that empirical demonstrations of psychology's unique ideological leanings are relatively few in number.

4

The Problem With Claims About Psychology's 'Liberal Bias'

Critics of psychology have drawn repeatedly on the same few studies when decrying the field's liberal bias (e.g., Bartels, 2023; Crawford & Jussim, 2018; Duarte et al., 2015; Haidt, 2016; Kaufman, 2021; Silander et al., 2020; Tetlock & Mitchell, 2015; Willis & Lancaster, 2020). Taking the reported findings as fact, they sweepingly denounce the field's toxicity and destructiveness, at times delineating elaborate psychological theories with which to explain the sheer virulence of psychology's cancerous partisanship (including, for example, the hypothesis that universities draw young psychologists into a spiral of 'self-radicalisation'; Honeycutt & Jussim, 2023). Given the alarming consequences implied by such extrapolations, it is important therefore to consider not just the scale and breadth of the evidence being presented, but also its robustness.

© The Author(s), under exclusive license to Springer Nature Switzerland AG 2025
B. M. Hughes, *Psychology's Quiet Conservatism*,
https://doi.org/10.1007/978-3-032-07724-0_4

Why the Empirical Evidence of Psychology's Liberal Bias is Weak

A frequently heard aphorism in behavioural science states that it is easy to conduct a survey, but hard to conduct a survey well. While there are very few barriers to prevent the production of survey datasets, there are equally few (and probably no) guarantees that any given dataset will be valid.

A plethora of complications can arise, each of which undermines the interpretability of survey responses. Survey takers may interpret questions differently to survey designers. Survey respondents may not report what they really feel, but instead respond in ways they think are socially desirable. Respondents may not have strong opinions about the subject matter, and so their answers might be more tentative than they appear. Survey questions are by nature verbal interactions, whereas respondents' *actual* behaviours and attitudes might differ from what they verbalise about themselves. And, of course, people who agree to take part in surveys might not be representative of the wider population that the survey taker, or anyone else, is interested in. Surveys are intended to find out what people think. But survey datasets—even the best ones—can only ever reveal 'what some people say about what they think they think' (Hughes, 2016, p. 154).

The fact that the evidence base for psychology's liberal bias amounts to a small handful of surveys should encourage caution rather than abandon. In this context, it is striking just how frequently and uncritically these surveys are actually cited, especially those conducted by Inbar and Lammers on members of the SPSP. The sample sizes for their two polls were 508 and 266 respondents respectively, with some overlap. Given that the SPSP had 7583 members at the time, this means that the surveys reached just 6.7% and 3.5% of their target population. This underwhelming coverage was partly explained by the fact that the surveys were sent only to members who had signed up to the SPSP's electronic mailing list. The problem was further compounded by apparent apathy among those who were contacted. With some 1939 members on the mailing list, the effective response rates for the two surveys were both very low, at 26% and 14%. When a survey invitation is ignored by six of every seven people who receive it, concerns immediately arise about the representativeness of the few who feel motivated to share their views. Before even considering the actual content of the dataset, it should be clear that the surveys provided little basis for sweeping conclusions. Claims they provided valuable insights into the thoughts of SPSP members—or of psychologists as a whole—should be seen as grandiose.

Added to this shortcoming is the fact that Inbar and Lammers's survey questions were not as clear-cut as might be assumed. A big problem is that their questionnaire was lopsided: respondents were asked lots of questions about conservatives but were not asked any questions about liberals. Similarly, the questionnaire focused on negative themes: respondents were not invited to share any positive attitudes they might have about conservatives or conservatism. This type of lopsidedness represents poor questionnaire design. It was inherent in the structure of Inbar and Lammers's survey that it would record some degree of anti-conservative sentiment, because it was not designed to measure anything else.

Deborah Prentice, a psychologist at Princeton University, drew attention to this imbalance in a follow-up paper published in the same issue of *Perspectives on Psychological Science* (Prentice, 2012). Prentice's paper also explained how the questions Inbar and Lammers used to inquire about psychologists' willingness to discriminate against conservative job applicants were ambiguously phrased. They asked about an 'inclination' to discriminate rather than an intention to actually do so. In other words, the psychologists were only asked about their emotional reactions in such situations. Whether they would ever act upon these reflexes was not probed. (Notably, Google Scholar shows that Inbar and Lammers's paper was cited nearly 600 times in the decade after it was published. By contrast, Prentice's paper identifying clear shortcomings in its methodology was cited just 27 times in the same period.)

Similar comments can be made about the other two membership surveys presented in Redding's chapter. The SPSP's own survey reached just fifteen per cent of its members, despite efforts to boost the response rate by extending the deadline and sending multiple reminder emails

(Society for Personality and Social Psychology, 2019). The survey of SESP members was based on just 335 responses (Buss & von Hippel, 2018), with around two-thirds of that society's membership not participating. Overall, while clear majorities of the samples self-labelled as 'liberal' rather than 'conservative', each sample contained just a small and non-randomly selected minority of the population being surveyed. The majority of the psychologists concerned did not express a view one way or another. Only a motivated minority had their opinions recorded.

Representativeness was also a problem with the other studies Redding cites, namely, the audits of US-based academic psychologists' registered political affiliations, conducted by Langbert and colleagues (Langbert et al., 2016; Langbert & Stevens, 2020). Redding explains that Langbert was unable to find Republican-registered psychologists in half of the country's top 40 universities, but he doesn't mention how many were working in the other half. Nor does he mention that the US is home to no fewer than 4000 universities in total, meaning that 'half of the top 40' comprises a tiny sliver of the US university system. Langbert's audit of the *four* highest ranking universities is even more hyperselective. Those four colleges not only represent a minute subset of all universities, they *by definition* lie at the extremes of American higher education. They are inordinately rarefied and utterly dissimilar to the vast majority of other degree-awarding institutions. Conclusions that are based on such bizarrely non-representative samples can hardly be seen as anything other than wholly unreliable.

It is notable that Redding's chapter makes no reference to the many published criticisms of the studies he describes. Nor does he offer his own assessment of the quality of the research. By presenting the findings without comment, he implies that they are valid on their face and that no critique is warranted. He makes no mention of contrary research findings, such as survey data suggesting that psychologists report higher levels of religiosity compared to other scientists (e.g., Delaney et al., 2013; Ecklund & Scheitle, 2007). He attempts no analysis of relevant controversies in psychology, such as concerns about the field's record of heteronormativity, masculinist bias, colonialism, ableism, or White-centredness, or its links with eugenics. He says nothing about the psychology curriculum's embeddedness in commercial publishing, or its adherence to market-driven priorities. Overall, Redding appears uninterested in the conventional academic balancing of pros and cons, of strengths and weaknesses, or of hypotheses and falsifications. In essence, he makes no effort to test his own argument against challenge. And why should he bother, given that, in his view, the truth of his claims is, ab initio, 'well understood'?

Such proneness to confirmation bias is especially ironic given the arguments advanced by Redding (and others) about psychology's supposed liberalism. Critics usually suggest that liberal psychologists cannot be trusted because they are willing to place their fingers on the scales of evidence. They are said to be guilty of scholarly partisanship: of drawing politically convenient conclusions from weak research by over-interpreting ambiguous data. And yet, when it turns out that the evidence cited to support the complaint is itself ambiguous and noisy, the critics suddenly overlook *any* methodological issues, before drawing conclusions that, lo and behold, entirely resonate with their own political worldviews.

As much as extraordinary claims require extraordinary evidence, it follows that ordinary evidence can only support claims that are, at best, ordinary. We might further conclude that *weak* evidence, if that is all that is available, should be seen as a red flag for claims that are weak.

Why the Anecdotal Evidence of Psychology's Liberal Bias is Questionable

Complaints about psychology's purported liberal bias do not rely exclusively on quantitative data. Some of the most strident are grounded in impression-based evaluations and anecdotal evidence. Those frequently recounted tales of campus cancel culture comprise their own body of anecdote, the shortcomings of which often stem from selective reporting. As outlined above, accounts of campus cancellations frequently omit important contextual details (such as cancelees' behind-the-scenes dalliances with political extremists). These errors of omission further discredit claims that it is somehow only *liberal* psychologists who play fast and loose with evidence.

Critics of liberal bias often accuse psychologists of 'abusing' language by formulating linguistic contrivances to distort reality in their favour. Phrases can be calibrated to convey a non-neutral or prejudicial meaning. For example, psychologists who study attitudes toward climate change often refer to 'climate change denial' (e.g., Cipriani et al., 2024). Some critics of liberal bias argue that this is unfair. They say that people who dispute the findings of environmental science should be seen as being *in disagreement* and not *in denial*. They argue that the concept of 'denial' implies a psychological state of disconnection from reality. Such language, these critics suggest, reveals a liberal bias because it implies that opponents of a particular political agenda are actually exhibiting a type of psychopathology (Duarte et al., 2015).

However, those critics are just as liable to do the same thing themselves. Heavily loaded euphemisms are frequently employed to cast implicit aspersions on 'woke' academics, whose chief crime is said to be, ironically, a failure to be fair-minded. For example, the US-based anti-woke advocacy group FIRE, the Foundation for Individual Rights and Expression (formerly the Foundation for Individual Rights in Education), arguably engages in propagandising wordplay *within its own name*. FIRE's main activities include campaigns to publicise incidents of campus cancel culture, as well as its online rankings of university wokeness. Harvard usually tops this latter list, with a free speech score of 0.00 and a 'Speech Climate' rating of 'Abysmal' (Gattuso, 2023). However, despite the appearance of empirical quantification, the scores on this database are compiled on an essentially qualitative basis. FIRE derives its numbers from survey ratings of attributes such as a college's 'tolerance for liberal speakers' and its 'openness'. The relevant survey for Harvard was based on a sample of 209 students, less than one per cent of the university's enrolment.

FIRE casts its mission as the protection of free expression. At first glance it appears as though it is fulfilling a worthy duty of defending the weakest members of society. But FIRE conflates its pursuit of free speech with a simultaneous objection to university policies that aim to promote on-campus diversity (Vivian, 2023). Their framing implies that initiatives intended to help socioeconomically marginalised students receive a university education somehow threaten the freedoms of those privileged enough not to require such support. As such, universities that attempt to address (societal) inequalities find themselves lambasted as the very *enemies* of equality, delinquent institutions failing to uphold the sacred values of 'individual rights and expression'. The terminology that FIRE employs appears to be deliberately emotive, a call-to-arms designed to whip up indignation in neutral audiences.

When criticisms are levelled at liberal psychology, vocabulary and terminology loom large. In another chapter of *Ideological and Political Bias in Psychology*, psychologist William T. O'Donohue of the University of Nevada (who, like Redding, serves as one of the volume's editors) sets out his personal critique of how 'social justice' agendas distort the neutrality of his field (O'Donohue, 2023). His examples include several concepts studied by psychologists that O'Donohue says fall short of scientific robustness. He offers a lengthy list of apparently offending language—including such terms as 'heteronormativity', 'homophobia', and 'ageism'. He contrasts these with vocabulary used in other sciences—such as 'hydrogen' and 'electron'—as well as with psychological terms that he declares to be scientifically acceptable—such as 'positive reinforcement' and 'conditional stimulus'.

Overall, it is unclear why exactly O'Donohue finds some terms more scientific than others. His unstated view appears to be that scientific entities should be more tangible, and thus more measurable, than others. It could be that O'Donohue feels that heteronormativity, homophobia, or ageism simply *do not really exist*, other than in the minds of (some) psychologists, and, as such, that they are projections rather than truly scientific constructs. However, it is well acknowledged in the philosophy of science that tangibility, in itself, does not make a construct scientific. A broad range of sciences study intangible constructs. Physicists study entities such as neutrinos that are difficult even to detect, never mind to measure. Epidemiologists infer the occurrence of important events from patterns in data. The idea that something must be materially objectified before it can be scientifically studied is just a fallacy.

In any event, for O'Donohue to make his case, he would need to demonstrate that heteronormativity, homophobia, and ageism do not in fact exist in the way that (liberal) psychologists claim. Unfortunately, although a co-editor of the 950-page book in which his own chapter appears, pressures of space seem to have prevented him from going further: '...it is beyond the scope of this chapter,' he points out, 'to examine each of these terms individually...' (p. 184).

Among the many other examples of concepts in psychology that O'Donohue says lack scientific robustness are 'inclusiveness', 'hegemony', 'sustainability', 'intersectionality', 'uncomfortable', 'offensive', 'climate justice', 'community', and 'safe space'. It is entirely unclear that any of these terms fall foul of the value-judgement that O'Donohue wishes to espouse. They are not even psychological terms. For example, when psychologists conduct research on the topic of 'climate justice', they are usually referring to the issue as it is discussed in general society. The phrase is not an invention, or a projection, of psychologists. Other items on O'Donohue's list of shame include 'LGBTQ+', 'Womxn', and, intriguingly, 'woke' (as if it were liberally biased psychologists who use this term or who are obsessed with defining or measuring such a thing). How these words can meaningfully be juxtaposed with 'hydrogen' or 'electron' seems pretty unclear. It is difficult to avoid the conclusion that O'Donohue's list merely contains a collection of buzzwords known to antagonise right-wing commentators, rather than anything genuinely representative of what researchers in psychology actually study in their work.

Anecdotal evidence is, by its very nature, untrustworthy. More than that, it is liable to be systematically misleading. Research has established that anecdotal accounts are not only often incomplete, they are frequently embellished by biases and faulty recollections. Anecdotes lead to generative misinformation. This is why hearsay is inadmissible in a court of law. In the history of forensics, psychologists have been at the forefront in highlighting its many shortcomings. It is therefore always notably ironic whenever psychologists mount arguments based on their own anecdotal accounts.

5

Culture War Psychology: Why the Liberal Bias Myth Persists, and Why It Is Damaging

Whenever psychologists' political opinions are surveyed, response rates are uniformly very low. This pattern might be said to represent its own form of testimony. When invited to express a view about their field's political bias, the vast majority of psychologists seem apathetic. Perhaps they are unmoved by politics in general, or wish to avoid polarisation. Or perhaps they decline to participate precisely because they believe such surveys are themselves politically motivated. Maybe they are suspicious of efforts to import the divisions of wider society into their professional domain.

The observation that psychologists are generally apolitical is consistent with research suggesting that most university academics actually identify as political moderates. While self-styled liberals certainly outnumber admitted conservatives, the largest group occupy a neutral zone between the two poles. These neutral professors are overlooked when political headcounts are binary, but they frequently comprise the plurality of opinion when surveys are designed to include them (Tyson & Oreskes, 2022).

Academic bias is exaggerated in the public imagination because controversies are highlighted disproportionately. Critics draw attention

© The Author(s), under exclusive license to Springer Nature Switzerland AG 2025
B. M. Hughes, *Psychology's Quiet Conservatism*,
https://doi.org/10.1007/978-3-032-07724-0_5

to cherry-picked examples for political reasons. In reality, only a tiny number of universities make the news. In the vast majority—literally tens of thousands of colleges—life just goes on pretty conventionally, with little signs of anybody revolting any time soon (Vivian, 2023).

Most universities are medium-to-large-sized corporations, operating within a globally capitalist market, developing strategic plans, appointing chief financial officers, recruiting marketing consultants, forever branding and rebranding themselves, all scrambling for an ever greater share of the end-of-the-rainbow pot of gold that is international student recruitment. In the main, modern universities are tools of neoliberalism—institutions necessary for training the next generation of capitalists, consumers, and climate wreckers.

The millions of students who pass through these places are subjected to endless corporate propaganda. They are bombarded with assurances that *their* university is best placed to guarantee them future career prosperity, an outcome that is typically indexed only by starting salary. Even the arts and humanities are spun out in ways intended to tantalise undergraduates (and their fee-paying parents) with seductive promises about the career-worthiness of 'digital' humanities, creative 'technologies', and arts 'management'. Most university business schools are committed to a capitalist worldview (Rhodes, 2022). Most university medical schools are shaped by medical commercialism, necessarily required to prepare their trainees for careers of entanglement with health insurers and Big Pharma (James, 2016). The research infrastructures of modern universities are obsessively calibrated to maximise 'success' in garnering grants from federal and supranational entities, in the process implicitly signalling that there will be no political dissent here—corporate universities do not customarily choose to bite the hands that feed them.

Our modern colleges' marketing departments tend to be coming down with no end of press officers, brochure designers, and advertising managers. By contrast, our counselling and student mental health departments are usually understaffed, shrinking toward disappearance as a result of staff burnout and managerially imposed fiscal austerity (Lightfoot, 2018).

Academics, for their part—with psychologists well and truly among them—play their part as corporate cannon fodder, keeping their heads down as the university proceeds to do its business. Precarious employment and the tenure system serve to keep them in their place. They are reduced to acting more or less as gig performers, reliant for status on subjugating themselves to a never ending series of metrics-based vanity exercises, pitching their intellectual labour to grant-awarding bodies and promotions committees alike, declaring their own ideas to be 'unprecedented', 'innovative', and of course 'excellent' (Vinkers et al., 2015), paying fealty to journal editors and anonymous peer reviewers ('*We thank the reviewer for their excellent observations; our paper is greatly improved as a result of their input*', etc.), with many modern-day faculty, it would appear, willing to step on heads in search of grander citation counts and ever higher *h*-indexes. Ironically, the faculty members who prosper most in this environment will be the very ones who end up leading it, usually then unsurprisingly choosing to reinforce its toxic norms.

There is little doubt that some psychologists—to repeat, *some* of them—are vocal on social justice matters, and actively pursue fundamental changes to what they feel are intolerable social conditions. But that is only some psychologists. Many psychologists, and possibly a large majority of them, are instinctively apolitical. With academics as a whole embedded in neoliberal corporate environments, it would seem that psychology is merely part and parcel of the Western status quo, quietly playing along with—and thus perpetuating—social conditions as they always have been.

So what then drives the view that psychology is emblematically liberal? Why do critics claim that a liberal mind virus is destroying the quality of psychological science? Why is it said that psychologists, as a group, are viciously partisan, engaging in a campaign of academic cleansing, cultishly hunting down and expelling conservatives from their midst?

In short, why is there always such a fuss about liberal bias in psychology?

Three Reasons Why the Liberal Bias Claim Persists

Just because there is a long-running discourse on psychology's liberal bias does not mean such a bias actually exists. Political controversies often turn out to be based on rumours rather than reality. There are at least three possible reasons why the *appearance* of liberal bias can persist even though the field of psychology, and most of its practitioners, are far from authentically liberal. Myths can thrive because of an undue focus on high-profile norm exceptions. They can last as a result of culturally idiosyncratic language connotations. And they can proliferate due to the political dividends of name-calling.

The first reason relates to the fact that true liberal bias is the exception rather than the norm. A key driver of *claims* about liberal bias is the fact that *examples* of liberal bias are themselves relatively rare. By their nature, however, exceptional cases stand out. This is because of what cognitive psychologists call the 'availability heuristic', a habit of reasoning to which all human beings are prone (Macleod & Campbell, 1992). The availability heuristic is where we attach special importance to easily remembered examples (or examples that are more 'available'), presuming them to be more significant than examples that do *not* spring so readily to mind. Hence, as classroom discussions tend to point out, we end up feeling fearful of air travel because of all those vivid news reports about plane crashes, but remain somewhat blasé about driving cars even though statistics show we are much more likely to die in road accidents. In much the same way, we think about bias in psychology in terms of a few headline-grabbing cases of academic liberalism, but fail to appreciate that most of psychology is, in reality, forgettably uncontroversial. It is inherent in the thinking styles of humans that we systematically over-estimate the prevalence of the unusual.

The availability heuristic compounds the problem of disproportionate media coverage. When attention is repeatedly drawn to events at Harvard, Cambridge, or similar universities, audiences would be forgiven for thinking that these institutions are typical of higher education as a whole, and that their specific crises provide generalisable lessons regarding all colleges.

In reality, however, they are highly peculiar places, accounting for just a sliver of the university system. These institutions stand out because they are, literally, exceptional: their goings-on are almost certainly *not* representative of higher education at large. Highlighting the experiences of so-called 'elite' institutions serves to mislead us about what the vast majority of universities, and their curricula, really involve.

Focusing on high-profile contentious events causes us to ignore the fact that most of what goes on in universities, even elite ones, is rigidly orthodox and generally uncontroversial. The majority of syllabi have little or nothing to say that would worry political traditionalists. Very few students of engineering, botany, mathematics, chemistry, computer science, aeronautics, equine studies, hotel management, dentistry, Latin, nanoscience, astrophysics, management information systems, or palaeontology will be encouraged in their classrooms to rise up and overthrow society. In fact, most course material goes unremarked upon precisely because it so closely coheres to society's more traditionalist norms. Those aforementioned business school curricula, for example, typically default to conservative market-oriented value assumptions when teaching students about pricing, competition, or social policy. Market mechanisms are framed as the only allowable social lever with which to ethically influence the world, as though collectivist action or higher taxation can have no good effect on our planet's well-being (Tyson & Oreskes, 2022).

Much the same can be said of the standard psychology curriculum. While it is absolutely the case that psychology students are taught many topics of applied policy relevance, most of their timetable is spent on material that is theoretical, methodological, reductive, or otherwise politically inconsequential. As fascinating as the minutiae of psychology can be, few revolutionary spirits will ever be stirred by classes on brain anatomy, signal detection theory, somatosensation, factor analysis, multiple regression, repeated-measures designs, item response theory, social facilitation, schemas, the differences between operative and figurative intelligence, the filter model of attention, or the notion that short-term memory operates like a Y-shaped tube. In reality, most actual psychology is of this genre. What critics such as O'Donohue discuss—concepts such as 'whiteness', 'Womxn', and so on—is simply not representative.

The second main reason why claims about liberal bias in psychology remain common despite being questionable relates to language. The notion of 'liberal' is inherently relative, referring as it does to a rather vague idea of whether people are comfortable with (or happy to permit) the prospect of change or difference. One person's understanding of what such change or difference might actually involve may not be the same as another's. Comfort zones, too, vary in their scope. Overall, it is difficult to precisely define what makes one person liberal and another person not. The term can be applied to a wide variety of targets. Complaining about liberal bias is easy when you do not have to explain what you mean.

This issue of vagueness is complicated by the way vocabularies emerge within different political communities. In the United States, where claims of psychology's liberal bias were first emphasised, the adjective 'liberal' has some very local connotations. The descriptor is employed pointedly in popular discourse as a non-specific term of abuse. American politics are exceptionally polarised (Heltzel & Laurin, 2020), and labelling is rife. The pejorative use of the term 'liberal' is a feature of this confrontational milieu. In European politics, it is common for mainstream political parties to have the word 'liberal' in their titles, such as Britain's Liberal Democrats. But in the US, candidates have been so afraid of being called 'liberal' that the term has at times become unmentionable. It has commonly been referred to as the 'dreaded "L" word' (Rosenthal, 1988)—the political doctrine that dare not speak its name.

The widespread practice of 'othering' opponents, exemplified by trolling strategies such as 'owning the libs' (Robertson, 2021), pushes politics away from the trading of ideas and into a realm of conquest, in which a candidate's electability is gauged not on their policies but on their skill at infuriating members of the other side (Finkel et al., 2020). Many American right-wing commentators will simply attach the adjective 'liberal' to whatever opinion (or person) they disagree with, without referencing any underlying political analysis or philosophy. While scholars who write papers on liberal bias in psychology will presumably have deeper ideas in mind, their use of the term 'liberal' functions essentially as a dog whistle. It quietly resonates with a wider constituency that lies far beyond academia, and which includes many in the right-wing media.

Given the potency of this tribal signalling, it is in retrospect unsurprising that even an ill-defined claim of 'liberal' bias in psychology would garner spiralling interest in the US context. In being so accused, psychology simply joins the entertainment industry, the legal profession, the technology sector, and a litany of other targets periodically attacked by American conservatives for being egregiously 'liberal'.

It is also worth recalling that what Americans consider 'liberal' might not be that liberal at all. Political scientists frequently note that mainstream American political manifestos contain policies that would, in most other countries, be seen as standard right-wing fare. While the Democratic party is not as right-leaning as the Republican party, it is by no means as left-leaning as it once was. What passes as 'liberal' in US politics currently represents, at best, a 'moderate' position on the global left-right spectrum (Hilbig & Moshagen, 2015). In short, by far the bulk of original commentary regarding the so-called 'liberal' bias in psychology emanates from a community that uses the term 'liberal' in a highly idiosyncratic way, arguably one that is quite divorced from its semantic (or psephological) meaning.

The final factor driving claims that psychology is liberally biased relates to the incentives underlying these political dynamics. Simply put, there are rich political rewards to be had from scapegoating 'liberal' psychologists, even if, largely speaking, accusations of such bias lack serious foundation. Well beyond the United States, social conservatives across Europe, the UK, Africa, South America, and Asia have learned that they will likely gain political ground whenever they accuse something—*anything*—of liberal bias. Psychologists make perfect targets because their field is both popular and socially relevant. Claiming that psychology is awash with liberal bias produces a political dividend for conservatives, ensuring that the practice becomes preserved within the political ecosystem.

Several factors make railing against wokeness in academia politically lucrative. Complaints that universities are riddled with social justice agendas will help anyone who wishes to discredit those very agendas in wider society. The accusation that academic staff—many of whose salaries are part-funded by taxpayers—are guilty of corruptly abandoning their professionalism adds grist to the mill for those whose political objectives include the defunding of public services. And let us not forget that sometimes students *do* acquire new ideas about the world when they learn something about what goes on in it. As such, a perpetual campaign of demeaning universities helps to demonise any dissent that emerges. What better way to dismiss young people's concerns about climate change or colonialism than by casting them as the deluded outpourings of a miserably *mis*educated generation?

Ultimately, there are several reasons why critics might complain of a liberal bias in psychology, other than the actual existence of such a bias. Claimed concerns about psychology's liberal bias are, to a large extent, contrived anxieties. They reflect synthetic paranoia—a manufactured crisis.

The Needle and the Damage Done

Antagonism toward academics can be addictive. The constant sniping, trolling, and needling of those who work in universities can often seem self-perpetuating, as though critics derive energy from the calling out of leftism. Social media platforms, where much of the associated moral panic is played out, have become zones of toxicity for many academic psychologists. Large-scale surveys suggest that around a third of academics routinely receive harassment online, with social scientists among those most likely to be victimised (Oksanen et al., 2022). Academics are often seen as soft targets. They are hamstrung by their academic freedom: if an academic complains about abuse they receive on the internet, they will quickly be lambasted by a battalion of anonymous trolls accusing them of the unscholarly crime of refusing to listen to contrary opinions. 'Universities are supposed to foster open debate and disagreement,' they will taunt. 'You belong in a creche, not a university!'

The blood sport of internet harassment is damaging both to those who receive it and to those who witness it. The abuse takes place publicly, in full view of students, colleagues, and the wider community. Academics who are attacked online report high levels of distress. They also experience feelings of professional isolation, perceiving that they receive little social support from their colleagues or employers (Oksanen et al., 2022). As is typically the case with social media, women and minorities are most likely to be trolled. When academics are harassed online, many recede to quieter spaces. As such, onlookers who might otherwise benefit from new knowledge are deprived of the opportunity to hear what academics have to say. The hostile environment created by online trolling—often framed as the *defence* of free expression—in fact has a largely chilling effect on academic speech.

There was once a time when what academics had to say would be completely ignored, or gently mocked. Professors were seen as liable to become wound up over trivia, and the more trivial the concern, the more wound up they would get. Back in the 1960s, sociologist Wallace S. Sayre encapsulated the issue by pointing out that academic arguments were reliably so bitter because 'the stakes are so low' (Shapiro, 2021). However, in the modern era, this has changed. Today, the issues about which academics argue are seldom trivial. The stakes are high. Psychologists are frequently hauled over the coals for their views on race, gender, and human suffering. Widespread efforts to discredit academic speech—for example, by declaring it to be automatically 'biased' simply by virtue of the fact that it *is* academic speech—serve to diminish the quality with which these important issues can ever be discussed in public.

In reality though, as outlined above, the majority of academics are neutral when it comes to politics. Most remain politically silent in just about every situation. A minority are vocal and left-leaning. But it is also true that another minority are vocal and *right*-leaning (witness Frisby et al.'s 950-page compendium of contributions decrying psychology's 'ideological and political bias'). And some academics are quite capable of giving as good as they get on social media platforms, engaging in trolling of their own (Tattersall, 2019). It is by no means unheard of for professors themselves to take to social media to complain about woke-ism, or to attack psychology's supposed liberal bias in vituperative terms.

The knee-jerk nature of anti-academic harassment necessarily elides nuance. Very few detailed debates can be had in fora where contributions are fragmented, cursory, and stylistically coarse, or, for that matter, where many of the most domineering voices are anonymous. The post/re-post/reply/like format has proven to be structurally polarising, and social media companies themselves appear motivated to preserve, rather than dampen, its outrage-provoking effects. Sweeping issues of existential controversy are reduced into meagre hashtags. Hashtag reasoning itself now regularly leaks into real world political contexts. Some of the most successful political movements of this century appear to have been premised on little more than catchy slogans, often of just three or four words (classics of the genre including 'Take Back Control', 'Get Brexit Done', 'Make America Great Again', and even 'Yes We Can').

It's Only Words

One of the many paradoxes produced by this reductionism is the perversion of ordinary language. Relentless repetition turns language into noise as listeners become desensitised to semantics. It has been some time now since the meaning of the phrase 'politically correct' was exhausted in this way. Opponents argued that, in this context at least, being 'correct' is in fact a *bad* thing, while being 'incorrect' would actually be *good*. Their unyielding campaign of repeating this charge at every available opportunity ultimately tainted the idea of 'correctness' itself, normalising the stigmatisation of what was previously an elementary concept, the ordinary meaning of which had been widely understood for centuries, even by children.

'Woke' itself is a modern counterpart of 'politically correct', and suffers from the same desensitisation effect. As is generally well known, the term alludes to the wisdom of staying 'awake' to the subtle and often taken-for-granted biases that lead to discrimination, injustice, and social exclusion. A 'woke' person is someone who is vigilant and attentive to how the world operates, and not someone whose mind is sleepy, sluggish, or dormant. And yet, just like the idea of 'correctness', the notion of being 'woke' is now dismissed by critics as a kind of character flaw. Through repetition and association, criticisms of the term have served to obscure its literal meaning. The implication now is that it would be better *not* to be attentive, that it would instead be advisable, presumably, to remain mentally asleep. (Ironically, many of these very same critics will just as stridently berate citizens for *not* thinking for themselves, calling on them to 'to do their own research' lest they become 'sheeple'; Hughes, 2023.)

In much the same manner, critics attack 'virtue signalling' as though the world would be a far better place if people were to signal their vices instead, or, at the very least, if they were to conceal their virtues. They decry anyone they can label as 'Antifa', as though being pro-fascism would be the morally superior position. And they see the aspiration for 'diversity, equity, and inclusion' as malevolent in and of itself, ignoring the literal meaning of the three nouns, and implying that the combined phrase signifies something actually abhorrent. When one liberal university president was forced to resign, right-wing media commentators in the US celebrated her removal as marking 'the beginning of the end for D.E.I. in America's institutions' (Confessore, 2024), as though all right-minded citizens should view diversity, equity, and inclusion as nefarious things to be hated and quashed wherever they might arise. With this reasoning, the removal of an antagonist from her job is one thing, but the actual 'end of equity' it signifies is the real benefit that should explicitly be heralded.

Breeding distrust in educators and undermining scientists are strategies that have been pursued by powerful interests for decades, if not for centuries—from the prosecution of Galileo for his heretical views on heliocentrism, which had threatened to destabilise the seventeenth-century church, to the concerted efforts of Big Tobacco, Big Food, Big Pharma, and Big Oil to ignore, suppress, manipulate, and misrepresent scientific research in their domains (Shulman, 2006). An endless drip-feed of name-calling, reputation-trashing, and obfuscation helps to shift the frame of reference whenever evidence is discussed in the public domain. Expedient politicians are enabled to impugn carefully researched reports as 'fake news' or to complain that data should be ignored because ordinary people have 'had enough of experts' (Mance, 2016).

Ultimately, the purpose of demeaning academics is to disempower them within the public square by casting doubt on their bona fides, by obscuring their ethics, and by generally making their lives miserable. Claims of scholarly bias, as well as labels such as 'liberal', help to achieve all three of these outcomes.

6

Liberal Bias in Psychology: An Intellectual Mirage

Psychology's 'liberal bias' narrative is as misleading as it is mis-informed. The field of psychology is invoked in many culture war contexts, but that merely reflects its relevance to those disputes rather than any partisanship among psychologists. In the end, the vast majority of voices heard in debates on gender identity, sexuality, race, colonialism, sexism, climate action, social justice, and so on, do not belong to psychologists. And the vast majority of psychologists never—*ever*—participate in these or any other political discussions, either publicly or in private. Rather, the myth of liberal bias in psychology is widely promulgated because claims of liberal bias in academia are *themselves* part of the culture war. And, as the cliché has it, the first casualty of any war is the truth.

This book aims to examine the actual truth of psychology's political contours. It focuses on the ways in which psychology has been shaped by conservative values, and on psychology's contributions to social conservatism both in the Western world and globally. It seeks to subvert the stereotype that psychology is an inherently liberal pursuit. Beyond merely linking psychology to historical or contemporary conservative movements, a key theme will be to show how psychology's

© The Author(s), under exclusive license to Springer Nature Switzerland AG 2025
B. M. Hughes, *Psychology's Quiet Conservatism*,
https://doi.org/10.1007/978-3-032-07724-0_6

focus on the individual—its stance of *individuocentrism*—itself promotes a politically conservative mindset.

In the coming chapters, we will examine the interplay between capitalism and human welfare. We will explore how the theories, research methods, and professional practices of psychology are designed for, and thus reinforce, a perspective that sees no alternative to a neoliberal capitalist economy. We will see how, as a result, psychologists deal with the human experience in terms of an individualistic, even bourgeois, worldview, in which emotional well-being is reduced to a set of computable variables, commodified products ready to be packaged for privileged consumers and delivered by commercial interests. We will consider the many ways in which scholarship in psychology reflects, and thereby rewards, traditional cultural value systems, such as heteronormativity. We will critique the presence of racism, sexism, and classism in psychology, and consider whether psychologists have succeeded in their attempts to excise traditional prejudices from their field's subject matter. We will examine how, and why, academic psychology continues to disproportionately focus on the needs and interests of Western audiences, and its reluctance to address, or even discuss, existential threats to humanity.

The one thing, however, that psychologists *do* appear to care about is psychology itself, in particular its standing as a prestigious endeavour (and thus, by reflection, their own standing as prestigious individuals). As a result, the structures of academic psychology are often inherently self-preserving, even where they are found to be flawed. While some psychologists are indeed committed to using psychology to advocate for social change, the majority adopt a stance of political quietism: keeping their academic heads down, publishing rather than perishing, and assiduously disengaging from the major social issues of the day. The university ecosystem of impact factors, research grants, and corporate rankings drive this academic behaviour, and we will consider why it is a deeply bad thing that few psychologists are ever rewarded for an achievement so banal as having tried to change the world for the better. The book culminates with a call for action—an agenda of positive steps that psychologists and citizens alike can take to help address the political shortcomings of this field.

The Past Before the Present

But before we can do any of these things, we will need to examine how all these various conditions came into being in the first place. While most psychologists consider their field to be a somewhat 'modern' science—be that a science of behaviour, cognition, or neurobiology—in many respects, the standard histories of psychology's founding reflect an exceptionalist creation myth rather than a true historical account (Hughes, 2023). When we look into the past, actual political truths can sometimes disappear beneath the official memories that the establishment chooses to promote for itself. Formal histories are often little more than propaganda. They are designed to *prevent* us from fully appreciating the past, rather than to help us understand the relevance of what took place.

In the history of psychology, we ceaselessly hear about Wundt, James, Freud, Pavlov, Watson, Skinner, Piaget, Milgram, Zimbardo, and the rest of them. But while such figures represent a certain diversity of intellectual approaches, we rarely step back and take time to consider how it happened that all these White men were able to become the vanguard of this supposedly new field. We seldom discuss how the psychological ideas produced by some of history's greatest civilisations ended up being supplanted by the Western, university-bound, grant-expending, impact-chasing, quantitative science that constitutes this thing called 'modern' psychology.

Psychology's history has endowed it with an inherently conservative overriding paradigm: that of focusing on individual experience at the expense of situational, social, or political context. This atomising of the human experience by psychologists—attaching primacy to the study of *factors* that *influence* the *behaviour* of *individuals*—serves to diminish accounts that emphasise societal, political, or cultural intricacies. In doing so, psychology's mainstream methods of knowledge production are geared to yield highly politicised subject matter. They shape explanations of people's lives that elide the role of external agents, and interpretations of social problems that are framed in terms of sufferers' own (in)capacities rather than the social forces that created them. In short, psychology is structurally set up in ways that promote system-justification, buck-passing, and victim-blaming as go-to explanatory narratives.

In the coming chapters, we will consider how this so-called 'liberal' psychology emerged from deeply socially conservative roots, and the impact this continues to have upon the field's activities today—embedding social conservatism within its modalities of research and practice. We will examine academic psychology's origins as a branch of religious philosophy, one that for many decades harboured a view of humanity that was skewed by Christian theology, universal patriarchy, European colonialism, White supremacy, and American segregationism, and we will trace its seamless passage into our modern era of social classism, poverty-indifference, and curricular pandering to ephemeral First World problems. We will also show how psychology's formation shaped conventions of scholarship that are politically skewed in ways to which psychologists themselves appear utterly blind, systems that just always happen to operate on behalf of cultural in-groups, concretising the marginalisation—and exclusion—of anyone whose interests do not align with the needs of mainstream privilege.

Psychology has demonstrated itself to be a science of the status quo, sharply bound by society's rationalisations for itself, rather than a truly objective endeavour of human self-examination. The claim that psychology is incorrigibly woke, that the field is somehow epitomised by dogmatic liberalism and anti-conservatism, could hardly be further from historical or contemporary truth.

Part II

Conservative Psychology, Past and Present

7

Legacy Conservatism

It wasn't that long ago that the American president derided an entire continent as comprising 'shithole countries'. In his view, citizens of such nations made undesirable immigrants—far less preferable than 'people from countries like Norway,' he opined (Turiel, 2020), or those from other similarly 'nice' places ('you know, like Denmark [or] Switzerland'; Pengelly, 2024a). In doing so, he invoked a doctrine that at one point comprised the consensus among academic psychologists—the discredited notion that psychological prowess is linked to human ethnicity. When the president doubled down by praising the 'good genes' inherited by Americans with Scandinavian ancestry, he was resurrecting a trope of Nordic superiority from early twentieth-century psychology. It became unmistakably clear that his anti-immigrant rhetoric was steeped in historical pseudoscience (Tensley, 2020). This presidential mainstreaming of eugenic prejudices illustrates just how potently psychology's troubling past can resurface in present-day politics.

It would be a mistake to think of the history of psychology as being relevant only to historians or psychologists. Psychology's history—like psychology itself—occupies neither a bubble nor a vacuum. Psychological ideas are not confined to the bubble of academia; they seep out into the wider world and influence how human beings perceive and talk about each other. And psychology's knowledge base is not vacuum-sealed in a way that protects it from outside events; rather, how psychologists have come to 'do' the science of mental life is itself influenced by what takes place in the world at large. Political bias in psychology is not, therefore, an illusion. The politics of the real world impacts upon psychology as much as psychology itself impacts upon the politics of the real world. But what is often overlooked is that the dominant political context shaping this impact is, and always has been, conservative.

Even the most provocative conservative pronouncements are usually derived from notions that, in the past, gained credibility from scientific endorsement. Take, for example, the fundamental issue of whether all our fellow human beings have the same worth as we do. Psychology has long tolerated the idea of a human hierarchy, in which some people—and, by default, some *groups* of people—are seen as less proficient than others. Members of outgroups have been portrayed as morally and mentally deviant, a practice of 'othering' that facilitates their ostracisation and neglect. The imprimatur of psychological science has frequently been invoked to justify intergroup hostility. Demonising stereotypes that, at one time or another, have constituted psychological orthodoxy continue to influence political discourse today.

Their impact can be seen whenever policy initiatives casually dehumanise the most vulnerable people in society. In 2024, when the United Kingdom government proposed an updated Criminal Justice Bill to criminalise 'rough sleepers', they justified their punitive approach to the homeless by characterising them in terms of their body odour. They proposed to empower police to arrest people judged to be emitting 'excessive ... smells' (Quinn, 2024). This focus on personal hygiene is rooted in a longstanding psychiatric stereotype that portrays social outcasts as responsible for their own misery. It ignores the circumstances that make it necessary for adult human beings to sleep on the streets in the first place—and, indeed, the dire mental health impacts of having to do so. Its precedent can be found in Victorian notions that conflated madness with physical dirtiness, thereby marking out the deviant, deeming them unsuited to modern society, and justifying their social exclusion. It is but one influential trope from psychology's history that has helped shape, and thus legitimise, a harshness in modern discourse.

As we will see in Part V, in the right circumstances, the stigmatising of difference can quickly be reconfigured into the pathologising of dissent. Tribal disagreements and political rivalries can be reframed as a clash of the inherently rational ('us') versus the hopelessly irrational ('them'). While governments are less inclined than they used to be to incarcerate their opponents in mental hospitals—a practice that flourished in the former Soviet Union (cf., van Voren, 2010)—modern politics retains its penchant for psychiatric stigmatisation.

In the aftermath of the UK's referendum on Brexit, for example, both factions railed against the other side's mental pathologies. Remainers and Leavers accused one another of holding 'delusional' fantasies, losing 'touch with reality', displaying tendencies towards 'self-harm', and exhibiting 'collective madness' (Hughes, 2019). Meanwhile, in the United States, partisan dissenters are frequently accused of suffering from 'Trump Derangement Syndrome' (Goldberg, 2015)—a condition that members of one state legislature attempted to have recognised in law (Hoffman, 2025)—while social liberalism is attributed to the effects of a 'woke mind virus' (Higgins, 2023). The rudimentary practice of declaring people mad simply because they think differently (i.e., from *you*) is emblematic of how psychologists approached mental illness for more than a century. It bolsters a type of political polarisation that renders all disputes intractable: after all, why bother negotiating with the other side if they are, in essence, unhinged?

Throughout its history, psychology has provided ample social permissions to dismiss nonconforming behaviour as unacceptable and aberrant. This is despite the fact that psychologists have long struggled—and probably to date have failed—to produce any satisfactory definition of what constitutes 'normal' human existence. Many of their earliest attempts to explain the difference between normal and abnormal behaviour have been thoroughly debunked or revised.

And yet their legacies continue to influence public debate by reinforcing convenient stereotypes. An obvious example is the presumption that 'abnormal' equates to 'uncommon'—if a person's behaviour diverges from what is usually seen, then it is presumed to indicate that the person is in fact mad. The problem, of course, is that not all unfamiliar behaviours will be symptoms of insanity. Mental health clinicians have long ago discarded eccentricity as grounds for psychiatric diagnosis. And yet the stereotype, so deeply rooted as it has been for decades, lives on.

The Problem of 'Individuocentrism'

A more pervasive legacy of psychology's history is its tendency toward elementary explanations. Its habit is to frame complex issues using simplified concepts. The approach echoes early psychologists' attempts to explain human behaviour under laboratory conditions, where rich, contextualised experience needed to be reduced to measurable variables that could be analysed under the metaphorical microscope. The process involves an over-investment in the 'individual' as the object of scrutiny. As we will see throughout this book, it is this orientation of *individuo-centrism*—the construing of psychology as a science that focuses on the perceptions, choices, and experiences *of individuals*—that explains, and drives, psychology's inherent conservatism.

Emphasising personal agency instead of societal, political, cultural, economic, or historical influences helps to promote a view that human outcomes are caused by people's own choices and actions. Such decontextualisation absolves wider society from responsibility for social problems.

A stark illustration of this can be found in the way psychologists discuss issues such as police brutality, in which violence is laced with layers of sociopolitical strife. Psychologists tend to discuss police brutality as being rooted in the cognition of individual (rogue) officers, an approach that obliterates the social and political dimensions of what is an almost intractable human problem. It ignores the role of structural racism in producing societal conditions that, in many countries, have allowed law enforcement agents to target ethnic minority citizens with relative impunity (Bowleg et al., 2022). When bodies such as the American Psychological Association issue public statements about police abuses, their language frequently reveals a cultural loyalty to society's most privileged citizens. Centuries of structural racism are displaced by euphemisms describing the 'difficult dynamics' that are said to exist between the police and minority groups (e.g., APA, 2014).

Psychology's emphasis on individuocentrism leaves it poorly equipped to explain—let alone address—societal injustices. Worse, it pushes psychology to reinforce unhelpful cultural stereotypes about who exactly is responsible for human suffering. Individuocentrism steers psychology's explanatory narratives toward socially conservative worldviews, and far away from anything that could be construed as 'liberally' biased.

The Master's Tools

By and large, psychologists themselves are oblivious to the politics of their paradigm. An important point that we will emphasise repeatedly on these pages is that while many psychologists may well consider themselves to be liberals, *this does not mean that psychology, as a field, is liberally biased.*

Ultimately, social conservatism is all about protecting the societal status quo. By concentrating on how *people* can change but *society* need not, individuocentrism feeds into this socially conservative agenda. Individuocentrism helps psychology to fulfil what the critical theorist Isaac Prilleltensky (1994) once referred to as its 'social reproductive functions' (p. 9). Psychologists just carry on, performing their duty as unwitting recyclers of societal norms.

Many psychologists regard themselves as having empathy with the vulnerable and excluded. But most are well and truly ensconced within society's privileged classes. Their lofty status is largely self-reinforcing. Those in a position to alter society's rules of engagement are its beneficiaries, and so usually have the least stake in actually doing so. Few influencers ever seek to undermine their own power to influence. As the philosopher and civil rights activist Audre Lorde once explained, 'The master's tools will never dismantle the master's house' (Lorde, 1984).

Professional psychologists, whether academics or practitioners, tend to lead sheltered existences. They are generally well looked after, well respected, and relatively well paid (university salaries are not what they once were, but they still provide tenured faculty with comfortable incomes). Tellingly, academics are not randomly selected from the wider population, but are chosen largely *on the basis* of enjoying pre-existing social privilege. Universities remain preserves of the already economically well off. Students from marginalised backgrounds are drastically under-represented in university classrooms, report experiences of discrimination and isolation while at college, and are far less likely than their peers to proceed to postgraduate study (Reay, 2018). Only those who survive this filtering process will become qualified academics. Unsurprisingly, therefore, most lecturers and professors will have benefited from social capital throughout their lives. Very few will ever have experienced genuine vulnerability or exclusion from their communities.

Still fewer will have experienced subjugation on a global scale. This raises the issue, which we will return to in Part VI, of psychology's domination by Euro-American thinking. The field's intellectual centre of gravity is the post-colonial West, a geopolitical entity that holds a disproportionate share of the world's economic power, military might, and political influence. The human experience described in psychology's textbooks and journals reflects the concerns of a very small subset of humans—primarily those who live in highly developed countries, whose basic health and sanitation needs are met from birth. As a result, psychology has become a science of First World problems. It trades in knowledge that applies primarily to the richest societies on the planet, whose populations in return hold the field in high esteem (cf., Moghaddam, 1987).

When we consider the history of the field, it should be no surprise to find that the type of psychology that has emerged in this milieu does very little to threaten the status quo. To do so would be to strive for the degradation of its own status. As a profession and a science, the psychology we have today is the type of psychology that prospers in the world exactly as things are now, culturally centred within humanity's most privileged societies, and with little obvious impetus to overthrow capitalism any time soon. Psychology is one of the master's tools. It is there to maintain, not bring down, the master's house.

8

What the Standard History of Psychology Usually Ignores

Even the way we talk about the 'history of psychology' is inherently conservative. A typical introductory textbook will describe the history of psychology as a series of epochs, involving contributions, by great men, of discrete paradigms of thought. Its narrative will begin with a founding event, set in 1879, when Wilhelm Wundt was said to have opened an 'institute for experimental psychology' at the University of Leipzig. Prior contributions from Ancient Greek, Roman, and Medieval philosophers will occasionally be packaged into a prologue, a kind of prehistory for the field (e.g., Robinson, 2013). Some such textbooks will counter that the area we today call 'psychology' simply did not exist before 1879 (c.f., Danziger, 2013). Others will argue that its modern incarnation is so particular as to require 'Psychology' to be spelled with a capital 'P' (e.g., Richards & Stenner, 2023), as though the field should be trademarked lest imitators get too many ideas.

From its initial Leipzig creation myth, psychology will usually then be described as flourishing in multiple forms. Loosely speaking, these will be grouped into a science-focused stream (comprising such movements as structuralism, functionalism, behaviourism, and Gestalt psychology) and an essentially unconnected person-focused stream (including, say, psychoanalysis and humanistic psychology). It will be suggested that both streams were eventually disrupted by successive technology-inspired 'revolutions', involving developments in areas such as cognitive psychology, neuroscience, and behavioural genetics. An important implication of this meta-narrative of constant progress is that these newer sub-specialisms, such as neuroscience, are presumed to be inherently superior to those that went before.

Of course, the history of psychology is a diverse field of study in its own right. Many scholars deviate fundamentally from the standard narrative. Several highlight the social and cultural factors that shaped the development of psychological ideas, and challenge the traditional 'great man' trope of historiography (e.g., Brysbaert & Rastle, 2021; Hughes, 2023; Jones & Elcock, 2001; Walsh et al., 2014). Nonetheless, the historical narratives and internalised cultural biases inherited by modern scholars remain somewhat overpowering. Few histories of psychology can avoid parading the same 'great' men when describing the field's past, even when criticising it. Histories remain rooted within the one constrained geographical perspective. Every historian of psychology mentions Wundt. Everyone mentions Freud. Everyone mentions William James. Few, if any, mention Xun Kuang, Girindrasekhar Bose, Frantz Fanon, Ibrahima Sow, or Oliva Espín. Even the most enlightened (or 'liberal') historians (e.g., Walsh et al., 2014) refer to psychology as having been 'imported' to Asian societies by 'Westerners' in the 1880s—as if Asian cultures had produced no indigenous analyses of human thought, emotion, or behaviour throughout the centuries that preceded their academic colonisation.

While it is apparent that many historians of psychology question the clichés of the standard narrative, it is less clear that they have overcome the psychological impulses that propel such thinking. Retrospection is inherently difficult. Even the more progressive scholarship exhibits apparent cognitive biases, such as hindsight rationalisation, in-group exceptionalism, out-group homogeneity, just-world thinking, survivor bias, affluence bias, moral relativism, presentism, internalism, and adumbrationism (Hughes, 2023). Standard histories of psychology often come across as celebratory rather than inquisitory, as though their main purpose is not to establish an awareness of psychology's past, but to socialise early-career psychologists into a protectionist academic guild (Lovett, 2006).

The way we discuss the history of psychology is shaped by our motivations and, by extension, our politics. In this sense, as twee as it may sound, the *psychology of history* is as important as the *history of psychology*. The conservatism inherent in the field can be seen from the way its practitioners present its past. It can be illuminating to reflect not just on what the standard historical narratives incorporate—but also on what they exclude.

9

Psychology's Roots in Theology

By presenting their field as the 'science of behaviour', psychologists ground their work in pure empiricism, far removed from superstition, mysticism, magic, or spirituality. The history of psychology is framed as part of the history of science itself. Psychology is portrayed as the product of Enlightenment values such as experimentalism, mathematisation, and the development of universal scientific laws. The triumph of psychology is presented as one of scientific reasoning over romantic religiosity.

It might therefore be presumed that the arrival of so-called 'modern' psychology into universities in the late nineteenth century was highly contentious, and that its appearance must have caused deep offence to what were widespread religious sensitivities. After all, at that time, the workings of human consciousness—the rudiments of the *soul*—were, for most of society, a very spiritual matter. Universities themselves were religious institutions. European and American colleges were akin to seminaries, and most were led by Christian clerics. Their curricula reflected a religious ethos. Theology and Latin were core subject matter, and students were expected to be conversant in the contents of the Bible.

© The Author(s), under exclusive license to Springer Nature Switzerland AG 2025
B. M. Hughes, *Psychology's Quiet Conservatism*,
https://doi.org/10.1007/978-3-032-07724-0_9

Science in universities was steeped in religion-friendly philosophy. The modern term 'natural sciences' reflects the ancient distinction between *natural* and *revealed* theology—the former exploring topics using reason, the latter drawing knowledge from the Bible, 'revealed' to humankind through scripture. 'Natural sciences', therefore, were originally those that could safely be explored without transgressing the boundaries of religious sanctity. Psychological topics could be included in such a curriculum, but only if deference was paid to the divine.

The subject matter of psychology was long present in the 19th-century curriculum, packaged as 'mental philosophy', a theologically-oriented branch of philosophical teaching. Drawing on British empiricism, which emphasised sensory experience and observation, the early mental philosophers emphasised the use of evidence in their work. They derived such evidence from introspection, anecdote, and observation, but also from truths 'revealed' by holy scripture (Fuchs, 2000). By the end of the century, dozens of textbooks had been written from this standpoint. A common curriculum for 'psychology' emerged, covering such topics as sensation, perception, memory, mental association, emotion, intellect, judgement, free will, attention, rationality, localisation of brain function, and even sometimes evolutionary theory. These were presented alongside coverage of more holy matters, such as 'the immortality of the soul' and 'the union of soul and body' (e.g., Maher, 1890). It was from this pious heritage that the field of psychology took shape in Western universities.

In reality, the arrival into academia of more scientific forms of psychology was not that disruptive. Tales of 'revolution' are exaggerated. On the contrary, by historical accounts, the emerging field was welcomed warmly by the religious university sector. Theologians saw little to fear from the 'new' forms of psychology, and in fact found much that stimulated their interests. Any form of study of the mind was viewed as important subject matter, because it shed further light on the nature of creation while illuminating the limits of human capacities to know and understand the world (Fuchs, 2000). The extensive catalogue of pre-existing mental philosophy textbooks allowed psychology courses to blend seamlessly into existing curricula. The key departure achieved by the new field was to quietly excise some of their more theological chapters, while *more or less retaining the rest of the established content.* Laboratories, of course, were also introduced, but these were generally seen as supplementing, rather than supplanting, the traditional approaches of introspection and observation (Dewey, 1884). The emergent field of psychology was viewed as a refreshment of existing offerings—a suite of new methods and instruments, rather than a threatening or interloping discipline.

And as the old theologians were comfortable with the new psychologists, it is important also to note that the new psychologists were, by and large, more than comfortable with theology. A large majority were deeply religious. The inaugural president of the newly established American Psychological Association, G. Stanley Hall, presented his field to the world as a religious vocation. Psychology was 'Christian to its root and centre,' he declared (Hall, 1885). He viewed the Bible as 'the greatest textbook of psychology' ever written (Hall, 1894), and hailed the human brain for serving as 'the mouthpiece of God' (Hall, 1901). Hall, who had sought a religious ministry prior to becoming an academic, peppered his writing with allusions to religious beliefs. He framed psychology as a moral endeavour aimed at character-building and spiritual enlightenment. His personal religiosity was mirrored in the vast majority of his contemporaries. Many joined Hall in adopting a proselytising approach, aggressively promoting psychology in the popular press as a complement to religious faith (Pickren, 2000; Fuller, 2006).

Even those early psychologists who were less avowedly religious were, in their own way, prone to piety and moralism. William James, who founded America's first psychology laboratory at Harvard in 1875, discussed with colleagues the prospect of using psychology to establish scientific proof for the immortality of souls. From this new science, James felt, psychologists could propose a new 'popular religion raised on the ruins of the old Christianity' (Coon, 1992). And in the relatively more secular United Kingdom, academic psychology still needed to ingratiate itself with prevailing spiritual beliefs. As noted by psychologist M. D. Vernon (1965), psychology was slow to develop in British universities because their academics maintained a 'vague belief' that 'there is something slightly indecent' about studying the soul. Care was needed, Vernon wrote, to acknowledge that the human spirit comprised 'mysteries which cannot be approached through any type of scientific investigation' (p. 212).

Far from displacing religion, academic psychology was essentially born from theology and groomed into mainstream academia by religious proponents. In fact, it is quite likely that without the religious sympathies of its practitioners, the new field of psychology would have found it impossible to infiltrate the university system. Psychology's compatibility with theological principles explains why it so quickly prospered in the late nineteenth and early twentieth centuries. By smoothly blending with religious discourse, the emerging psychology supported modernising societies to embark on an incremental form of secularism, to gradually reformulate their spiritual outlooks in a way that accommodated scientific ideas, without ever having to fully discard the consolations of mysticism.

10

Psychology's Roots in Class Conflict

It is easy to overlook the fact that, when viewed in historical perspective, concepts such as universal suffrage, human rights, and equality before the law are quite novel, if not absolutely strange. While democracy has existed as a concept for thousands of years, it has existed as a practice for a much shorter period of time. For most of human history, it was considered entirely reasonable to believe that, when it comes to judging the value of human beings, some people were simply less equal than others.

Even when democracy is nominally introduced, it can take many decades for it to fully roll out. The French revolution of 1789 is often heralded as a milestone in democratic history, but French women had to wait until 1945 before being allowed to vote. Switzerland is said to have become a democracy in 1848, and yet the country's Jewish population was deprived of full political rights until 1866, while Swiss women were not allowed to vote in national elections until 1971. In one Swiss canton, the tiny Appenzell Innerhoden, women were deprived of voting rights until as recently as 1990.

When psychology was emerging as a formal university discipline, the equality of humankind was far from widely accepted. As William James

© The Author(s), under exclusive license to Springer Nature Switzerland AG 2025
B. M. Hughes, *Psychology's Quiet Conservatism*,
https://doi.org/10.1007/978-3-032-07724-0_10

started his career at Harvard, it was considered entirely legitimate in America to exclude Black citizens from civic life, with several states operating convoluted rules requiring literacy tests and tax certifications that were explicitly designed to disenfranchise African Americans. Things were much the same in Europe. When Wundt opened his psychology institute in Leipzig, the supposedly democratic German Empire still did not allow women to vote in elections. At the time Freud published *The Interpretation of Dreams*, Jews were banned from working in Vienna's local government. When a group of ten British academics gathered in London in 1901 to establish what would become the British Psychological Society, voting rights in UK parliamentary elections were restricted to a small minority of the adult population, namely, men who owned property.

The pioneers of academic psychology lived in a world in which social inequity was normalised. Endless foreign wars served as a backdrop to domestic turmoil. Scholars and scientists belonged to the privileged classes, and their musings frequently revealed paranoia regarding the threat posed by the 'masses'. The assumption that human beings should be stratified by social class was endemic in the societies in which the newly bureaucratised form of psychology first emerged. It was one of the working principles on which the modern academic discipline of psychology was constructed.

Various strands of psychological thought had speculated vituperatively about the proclivities of the poor. Cesare Lombroso, the founder of modern criminology, mainstreamed the idea that criminality was a heritable human trait that left physical marks upon the species. According to Lombroso, 'born criminals' could be identified based on their physical appearance. He argued that the faces of criminals resembled those of poor people, foreigners, and, indeed, subhuman primates, whereas those of the privileged classes displayed immutable virtue. This logic of physiognomy—the inference of psychological traits from physical appearance—was influential across psychology (Collins, 1999). Even Alfred Binet and Théodore Simon, developers of the first practical IQ test for children, were convinced that teachers could better evaluate a child's innate intelligence by examining their physical features. They claimed that 'the countenance is scarcely deceptive for those who

are used to reading it' (Binet & Simon, 1916). In their view, psychometric tests were just a standardised means by which to replicate this
experienced classroom judgement.

According to mainstream scholars, the poor posed a particular threat
when gathered as a group or 'mob'. The standard view was that crowds
exhibited an inherent proneness to chaos. This was perhaps an unsurprising conclusion given the frequency with which political uprisings
occurred at the time, a period that some scholars came to refer to as 'the
age of the crowd' (Moscovici, 1981). Influential European theorists such
as Gustav Le Bon, Gabriel Tarde, and Scipio Sighele depicted crowds as
vehicles of mass fury. Inherent in their thinking was the view that some
of their fellow citizens were incorrigibly savage, impossible ever to fully
tame through civilisation.

Le Bon argued that members of mobs would regress to earlier stages of
natural evolution, which, he said, accounted for their barbarism. He
based this conclusion on a baroque blend of classism and misogyny. In
Le Bon's view, the impulsivity, emotionality, and a lack of reason that
crowds displayed typified inherently feminine instincts. This, he felt, was
evidence of evolutionary regression, because, in his view, women represented 'the most inferior forms of human evolution', possessing brains
similar to 'those of gorillas' (Le Bon, 1879).

Tarde, one of the first scholars to publish his work under the classification of 'social psychology', felt that congregating in crowds encouraged people to engage in mindless imitation. In particular, he felt that
mobs caused human beings to swarm like insects, leading their members
to lose their individual minds. Sighele was one of those who called for
a specific branch of psychology to study the adverse impacts of social
congregation on human cognition. He was deeply sceptical of just about
any form of human grouping, or any system that relied on group-based
collusion. He argued vehemently against trial-by-jury and, indeed,
against the introduction of democracy (Jahoda, 2007).

A common nineteenth-century trope was to warn of the looming
catastrophe of species-wide human degeneration. A widely held view was
that humanity had passed its evolutionary tipping point, and was now
destined to return to a state of primitive savagery. The unpredictable and
uncivil masses were said to be early exemplars of this inevitable slide.

Across Europe and North America, a shrinking bourgeois elite found themselves surrounded by a restless populace. They saw psychopathology as the reason for the rising tide of civil agitation, setting the tone for how psychologists would come to view mental illness and its diagnosis. Increased provision of mental asylums in Europe and North America coincided with an expansion of scholarship in psychiatry. It was during this period that theorists began to formally classify social outcasts as mentally disordered.

In one particularly infamous example, an American physician concluded that African people who were unwilling to be subjugated as slaves were in fact insane. He declared that they suffered from 'drapetomania', a psychiatric condition purportedly characterised by the recurring impulse to flee enslavement (Bynum, 2000; Guthrie, 2004). The broader idea that slavery could be justified on the grounds of inherent psychological (and physical) differences between Black and non-Black people underpinned several durable racist stereotypes that were to leave their mark on American scholarship, and society, for decades (Deutsch, 1944).

Academic psychology expanded rapidly between the 1870s and the 1930s, a period characterised by global economic recessions, bloody and pointless wars, and widespread social malaise. All these forces propelled degenerationist anxieties. Psychologists openly discussed what they declared to be an epidemic of mental illness—not *caused* by stressors in the environment, but one that *accounted for* society's very demise. In 1881, the neurologist George M. Beard, in his book *American Nervousness: Its Causes and Consequences*, warned readers of the dangers inherent in steam power, the periodical press, and the telegraph (as well as in 'the mental activity of women'). Tarde blamed daily newspapers for destabilising the masses by exposing readers to the influence of nameless agents operating from a distance, becoming perhaps the first commentator to raise the alarm about remote radicalisation.

Prominent psychologists voiced strong concerns about the degeneration of the human species. Emil Kraepelin, a disciple of Wundt who is today acclaimed for his work in classifying schizophrenia and co-discovering Alzheimer's disease, was scathing of the lower classes. He bemoaned them as a proliferation of 'imbeciles, epileptics, psychopaths, criminals, prostitutes and vagabonds [who] are the children of alcoholic and syphilitic parents' and openly advocated for their physical removal from society. He campaigned against Germany's 'ever-expanding social welfare programmes' on the grounds that preventing the poor from starving to death would have the effect of 'impeding the natural self-purification of our people' (Kraepelin, 1908).

In Britain, the renowned English psychologist Charles Spearman echoed these sentiments. The poor, he felt, could not be trusted to contribute rationally to the world around them. He argued that the lower classes should be psychologically screened before being allowed to vote, or, for that matter, before being granted 'the right to have off-spring' (Hart & Spearman, 1912). Calls for the mass compulsory sterilisation of the poor were not at all uncommon during psychology's formative years.

For well more than a century, academic psychologists forged a curriculum of knowledge that reified the notion of a human under-class. The field grew in popularity in no small part because of its resonance with bourgeois classism. As we will see in Part III, this dynamic continues to leave its mark today. Individuocentrism in psychology continues to prime narratives that attribute poor people's mar-ginalisation to their inherent lack of initiative, rather than to their persistent oppression at the hands of the wealthy and powerful (Williams & Cabiles, 2016). Today's psychologists continue to pay infrequent attention to social class as a component of identity (Manstead, 2018).

Psychology's decades-long development as a field of social privilege, one practised by the rich and dismissive of the poor, is rarely acknowl-edged in psychology's official histories. It is as though poverty, and the poor, have been rendered invisible.

11

Psychology's Roots in Eugenics

Theology, classism, and degenerationism are often intertwined. Their confluence is apparent in another force that shaped the early development of scientific psychology—namely, the eugenics movement. Proponents of eugenics claim that the 'genetic quality' of the human population can—and indeed should—be improved through science. They discuss their social engineering objectives with reference to 'selective breeding', as though the human species can be refashioned in the same way that farmers cultivate more profitable varieties of cattle or wheat.

Doctrinal eugenicists have at various times promoted extreme interventionist practices to remove 'inferior' members from humankind. They have called for targeted abortions, compulsory sterilisations, enforced segregation, laws prohibiting 'miscegenation', and immigration policies aimed at excluding purportedly 'undesirable' outsiders. Today the eugenics movement is perhaps most infamously associated with the genocidal activities of Nazi Germany, where the Nazi regime massacred millions of people in the name of racial hygiene. Not without cause, the entire enterprise of eugenics has become notorious over time.

© The Author(s), under exclusive license to Springer Nature Switzerland AG 2025
B. M. Hughes, *Psychology's Quiet Conservatism*,
https://doi.org/10.1007/978-3-032-07724-0_11

The idea of selectively breeding human populations had been floated for many centuries, by philosophers, ethicists, and legal theorists. However, it was a psychologist who first brought the modern eugenics movement into existence. In his 1883 book, *Inquiries into Human Faculty and Its Development*, English psychometrician Francis Galton explained his long-standing belief that successful people inherited their talents from their parents. Reflecting on how this knowledge could be deployed for the betterment of humanity, he proposed a new field that would allow scientists to 'give to the more suitable races or strains of blood a better chance of prevailing' (p. 25).

At first he thought of naming the field 'viriculture', a play on language that invoked the idea of using agricultural methods to enhance the human (or virile) population. But Galton's vision extended beyond the application of agricultural methods. He felt that efforts to improve humanity should be 'by no means confined to questions of judicious mating' (p. 25), apparently acknowledging that efforts to curtail the spread of undesirable genes would require more than family planning advice. He therefore arrived at the more general name of 'eugenics'. This, he felt, was a briefer and more elegant word. It would be better suited to describing his new 'science of improving stock' (that is, the stock of human genes).

Galton was no fly-by-night grifter. He was an illustrious and influential psychologist, often lauded as the founding father of psychometrics. He developed statistical tools and procedures that are still widely used today, such as correlation coefficients, standard deviations, log-normal distributions, and scattergrams. Widely sought out as a mentor, Galton helped to launch the careers of many psychologists who themselves would go on to be acclaimed, including statistical pioneers whose names are now synonymous with some of the most popular techniques used to analyse psychology's research data.

One of his most famous protégés was Karl Pearson, who was a precocious statistician. Pearson not only devised the product-moment correlation coefficient (or Pearson's r), he also developed chi-squared tests, principal components analysis, histograms, probability distributions, negative correlations, contingency tables, and concepts such as random error. Another prominent follower was Ronald Fisher, who gave us analyses of variance (with their F-ratios), as well as Fisher's exact tests, z-distributions, random effects models, and null hypotheses. Galton, Pearson, and Fisher self-identified as psychologists, but their primary purpose in developing these methods was to further the field of eugenics. They wanted to make it easier to demonstrate the principles of eugenics using data.

The three men were unabashed in their support for eugenics as a science, and for all of its associated ideology. Galton travelled extensively in Africa, where he developed shocking views about the tribal communities he encountered. He felt that Black people were so inferior to Europeans that slavery would be a *good* thing for them to experience. Being 'under the hand of a slave-owner,' Galton wrote, would make African people 'less mischievous' (Galton, 1853, p. 140). He added that slavery would improve their lives, because they 'could hardly become more wretched than they now are' (p. 140). Pearson was an early proponent of what is now known as the 'White replacement' conspiracy theory. He worried that White people would soon be outnumbered by people of other ethnicities. 'When the white man and the dark shall share the soil between them,' he wrote, 'mankind will no longer progress; there will be nothing to check the fertility of inferior stock' (Pearson, 1901, p. 24). For his part, Fisher sympathised with any regime that attempted, by whatever means, to reverse such demographic trends. He was explicitly sympathetic toward the objectives of the Holocaust. In his view, the Nazis' aim 'to benefit the German racial stock' was worthy, and justified 'the elimination of manifest defectives' (as quoted by Weiss, 2010, p. 745).

Until relatively recently, Galton, Pearson, and Fisher were widely lauded as pioneers in the history of scientific psychology, figures to whom all modern psychologists owed debts of gratitude. Galton received an honorary degree from Cambridge University and was knighted by the king in 1909. Fisher too was knighted, while Pearson was offered a knighthood but turned it down. Lecture theatres named after Galton and Pearson were established at University College London, which also named an entire building after Pearson. UCL also named two professorial chairs after the men: the Galton Chair of Eugenics (later the Galton Chair of Genetics) and the R. A. Fisher Chair in Statistical Genetics. As recently as 2010, UCL established the R. A. Fisher Centre for Computational Biology. Fisher was also commemorated by a stained-glass window at Cambridge University. All these accolades were eventually withdrawn, but not until 2020, in the aftermath of global anti-racism protests prompted by the murder of an unarmed man, George Floyd, by a police officer in the United States.

It is sometimes argued that we should not judge historical figures by today's moral standards, as though morality exists outside history itself. The defence is that figures such as Galton, Pearson, and Fisher were simply immersed in the morality of their times. They lived in a society in which it was perfectly acceptable to hold views that would today seem profane. In other words, hindsight is a wonderful thing and these men simply did not know any better.

However, it is misleading to suggest that the eugenicists were oblivious to criticisms of their ideas. Although eugenics drew support from across the political spectrum (Rutherford, 2022), many high-profile detractors expressed abhorrence at its inherently racist, classist, sexist, and xenophobic assumptions. One prominent celebrity critic was the British writer G. K. Chesterton, who condemned the field in a bestselling book, *Eugenics and Other Evils* (Chesterton, 1922). Leading scientists ridiculed the fundamental premises of eugenics, including its assumptions regarding the heritability of traits. One was the American geneticist Thomas Hunt Morgan, who was so esteemed that he would later win a Nobel prize (Glass, 1986). In short, hindsight does little to exculpate Galton or his followers. They knew perfectly well that their ideas were regarded as scientifically contentious and morally repugnant. They just insisted on promoting them anyway.

It is notable that eugenics was—and still is—defended by its acolytes despite several scientific shortcomings with its paradigm. For one thing, the type of human social engineering envisaged by eugenics is entirely fantastical in real terms (Curtis & Balloux, 2020). Humans are not bred in controlled environments (like, say, dogs or cattle) and so it is impossible to identify who among us possesses, from birth, the 'desirable' traits that eugenicists feel we should target. Secondly, even if we were to rear humans in cages, they would still undergo long generational cycles and produce relatively few offspring, meaning that attempts at selective breeding would take literally centuries to have any effect. And thirdly, the link between genes and psychological attributes is just not as straightforward as the eugenicists claim. Most human traits are 'emergent' properties, meaning that they involve multiple genetic components and so are not amenable to selective breeding (Mitchell, 2018). Many are the result of *de novo* genetic mutations, which by definition cannot be predicted. And others are produced by recessively acting gene variants, which most of the time are entirely harmless.

All told, eugenicists might feel that their field is a science, but it lacks any meaningful scientific rigour. Eugenics, in reality, is a pseudoscience.

Eugenics as Mainstream

The eugenics movement was central to psychology throughout its formative years, and exerted a lasting influence over the field. It would be no exaggeration to state that, for several decades, eugenics and psychology were effectively inseparable. Some of psychology's most acclaimed figures were vociferous eugenicists. They repeatedly echoed and emphasised eugenic principles in their thinking.

The roster of presidents of the American Psychological Association offers a convenient roll call of cases. The aforementioned G. Stanley Hall, the APA's founding president, heralded eugenics as 'larger than pedagogy, religion and all other culture influences combined' (Hall, 1911, p. 152). Consistent with his religious views of psychology, he believed eugenics to be 'a legitimate new interpretation of our Christianity' (p. 157). Edward Thorndike, who became APA president in 1912 and is remembered as the founder of functionalist psychology, claimed that 'selective breeding' was needed for the human population 'to keep sane, to cherish justice or to be happy' (Thorndike, 1913, p. 13). Robert Yerkes, who served as APA president in 1917, was appointed an 'expert eugenic agent' for the US government. He testified that 'no one of us as a citizen can afford to ignore the menace of race deterioration' (Yerkes, 1923, p. viii).

Another APA president, Lewis Terman, is today acclaimed as the developer of the Stanford-Binet Intelligence Scales. However, his aims in devising a standardised intelligence test for Americans were far from humanitarian. Terman claimed that Hispanic and Black Americans were afflicted by a 'dullness' that 'seems to be racial, or at least inherent in the family stocks from which they come' (Terman, 1916, p. 91). He was willing to acknowledge that the eugenics movement was controversial, writing: 'There is no possibility at present of convincing society that [ethnic minority persons] should not be allowed to reproduce' (p. 92). Nonetheless, Terman felt that something needed to be done about such people. 'From a eugenic point of view,' he argued, '[ethnic minorities] constitute a grave problem because of their unusually prolific breeding' (p. 92).

In all, between 1892 and 1947, some 31 presidents of the American Psychological Association were publicly listed as leaders of formal eugenics societies or campaigns, with several others publishing papers supportive of the eugenics movement (Yakushko, 2019).

The problem here is not so much that these presidents held positions of influence. Rather, the problem is that they were *elected* into their influential positions because of what they stood for. Their strident supremacism reflected an outlook that was widely shared by psychologists at the time. Psychologists chose as their leaders unapologetic advocates for human sterilisation, segregation, and racial purity. They selected these men precisely *because* of their doctrinal leanings. They saw their ideas as representing the best of what their newly burgeoning academic field had to offer.

In the second half of the twentieth century, explicit support for eugenics became much less common once the horrors of the Holocaust were revealed. The word 'eugenics' almost (but not entirely) disappeared, but the shift occurred in vocabulary rather than ethos. Eugenic *ideas* remained influential. Some of the world's most prominent psychologists continued to profess eugenic positions.

Shortly after the war, the leading English psychologist Cyril Burt (himself a former president of the British Psychological Society) warned the British public that the birth rate among people of low intelligence in the UK was a problem 'large enough to demand urgent practical attention' (Burt, 1952, p. 5). He strongly doubted that 'the more fertile classes, though less intelligent, might possess other qualities' (p. 31), such as physical strength, that could be seen as useful to society. Instead, he foreshadowed the dangers of failing to sterilise the poor. Burt believed that unfettered reproduction by the lower social classes would produce a decline in national intelligence that 'would have grave effects on the mental status of the population if at all prolonged' (p. 31). In other words, he felt that something needed to be done to prevent the poor from having children.

Hans J. Eysenck, widely considered to be among the most illustrious British psychologists ever, persistently claimed that different ethnic groups differed in intelligence. Perhaps his most ominous writing on the subject was contained in his 1973 book, *The Inequality of Man*. In that book, Eysenck railed against affirmative action programmes in universities on the basis of what he believed was the 'biological inequality' between races. 'You can appoint a minority member to a professorial chair,' he wrote, 'but you cannot make him equal in ability and performance' (p. 229). Eysenck was thus an early proponent of the war on DEI in higher education. He was also a prolific writer and public speaker, regularly to be found defending other psychologists whenever they were accused of racism.

One colleague Eysenck frequently spoke up for was the British-American psychologist, Raymond B. Cattell. A giant of twentieth-century psychology, Cattell became renowned for his theories about human personality. Among other lasting contributions, he developed the landmark '16PF' model of personality structure. Cattell never trained in the biological sciences, but he nonetheless constructed elaborate physiological explanations for his hypotheses. For example, he frequently compared the breeding of human beings to that of plants. He wrote that 'the first requirement in successful plant hybridization is a rejection of perhaps 90% of the hybrids as unsuccessful,' before using this supposed law of organic self-destruction to explain how racial 'interbreeding' accounts for 'higher crime and insanity rates' among immigrants to the United States (Cattell, 1987, p. 202).

Cattell was forthrightly blasé on the matter of genocide. 'Nature constantly commits both homicide and genocide,' he wrote, 'and there is no question that both individuals *and races* are born to die' (Cattell, 1972, p. 220; italics added). He accepted that genocide was controversial, but felt that this was mostly because the word itself was commonly 'bandied about as a propaganda term' (p. 220). (Were he alive today, he might well have derided anti-genocide campaigners for being 'woke'.) Reluctant to trigger the libs, Cattell instead coined a new term—*genthanasia*—to refer to how measures such as compulsory sterilisation could be used to 'phase out' undesirable races, 'without a single member dying before his time' (Cattell, 1972, p. 221). But regardless of Cattell's evasive wordplay, such programmes of mass sterilisation still constitute genocide under the standard legal definition.

Inherited Eugenics

The mainstream textbooks on the history of psychology rarely focus on the field's century-long immersion in eugenics (Yakushko, 2019). Perhaps it is for this reason that up to the present day, a drumbeat of eugenics apologism appears to rumble on across the field. Some extremist psychologists persist in spouting racist theories in the darkest corners of the internet. But occasionally, ideas relating to selective breeding, genetic differences, and the innateness of human hierarchies continue to be heard within the mainstream of the field. It can sometimes be difficult to know for sure whether a particular psychologist seriously endorses the principles of eugenics, or whether they have just obliviously internalised aspects of its thinking, as if by osmosis.

Take for example the Canadian-American psychologist Steven Pinker, one of the foremost public intellectuals in modern academia. Pinker has written several books explaining psychology to general audiences. He is widely regarded as a supremely talented communicator and is known for his theories exploring the impact of rationality on the development of human civilisation. But Pinker is also known for drawing political inferences from psychology, and for being sceptical of social liberalism, especially as it arises in universities. He is a prominent voice criticising the purported 'liberal' bias in psychology and regularly rails against campus cancel culture. He routinely mocks liberals, dismissing them as 'leftists', or, more ornately, as proponents of 'PC[political correctness]-SJW[social justice warriors]-Critical-Woke -Intersectionality' (e.g., Pinker, 2025).

One of his many best-selling books, *Enlightenment Now*, celebrates the impact of Western culture on world history (Pinker, 2018). In *Enlightenment Now*, Pinker condemns what he frames as the political left's critique of cultural colonialism. He proposes that liberals should praise such colonialism rather than reject it. He asserts that the human population is now safer, healthier, and happier than ever before, and attributes this to the modernising effect of the scientific rationality that emerged in Europe during the seventeenth and eighteenth centuries. A key thrust of Pinker's case is that rationalism is core to traditional Western civilisation. From Pinker's perspective, the entire world should thank Western culture for the wonderful lives they enjoy today. He (rightly) complains about anti-science sentiment in political debate, drawing attention to the 'stupidification of science in political discourse' (p. 387) when politicians debate such issues as evolution and climate change. However, he is most concerned about an 'even deeper' hostility to science that resides in the halls of academia, amongst liberal intellectuals—on the academic left—who build their careers on anti-science scholarship.

As part of these reflections, Pinker considers the history of eugenics. He explicitly castigates scientific racism as a social ill. However, he nonetheless implies that criticisms of eugenics have been excessive. The eugenics movement, he says, 'has been used as an ideological blunderbuss' (p. 399), arguing that eugenics 'deserves a deeper and more contextualized understanding than [its] current role as anti-science propaganda' (p. 399). He suggests that the political ramifications of eugenics have been allowed to overshadow its rationalistic underpinnings. Pinker complains that the eugenics movement has become 'permanently discredited by its association with Nazism' (p. 399), a linguistic coding oddly reminiscent of Cattell's thoughts on how the term 'genocide' lost its value through political over-use. Overall, it seems that Pinker feels a key point to make about eugenics is that its underlying principles have been unfairly maligned by irrational, anti-science ('woke'?) voices on the political left.

Pinker attempts to bolster his case by mentioning worthwhile practices that he says have been curtailed because of unwarranted concerns about eugenics. He gives the example of how critics stigmatise the use of genetic screening in medicine, even when it is used to detect foetal abnormalities. With this argument, he appears to imply that (left-leaning) anti-science propagandists are so het up in their hatred of eugenics that they are willing for babies to be born with degenerative diseases—a take that conspicuously fails to acknowledge the broader criticisms of prenatal diagnostic screening, which encompass scientific as well as ethical concerns (e.g., Stapleton, 2017).

Eugenics is usually considered an obsession of the political right, and indeed of the Far Right. But Pinker attempts to argue that the eugenics movement is politically *left*-leaning and anti-conservative. He notes that eugenicists favour central planning and social reform, both of which are signature policy tools of the liberal left. He argues that the practical aims of eugenics require large-scale state intervention, an archetypally big-government approach that invokes left-wing notions of social engineering. Eugenics, he suggests, is the antithesis of *laissez-faire* individualism. The problem with this assessment is that it focuses on policy implementation rather than on policy per se. Eugenicists do indeed advocate for government-level interventionism—but then so do right-wing authoritarian dictatorships. Wishing that the state exert control over people's lives is an unconvincing basis on which to impugn a movement for having 'leftist' leanings. What is more important to consider are the objectives such activism is intended to pursue.

What Pinker fails to acknowledge is that the assumptions of eugenics are grounded in ideas that epitomise right-leaning political worldviews. Eugenics offers an entirely neoliberal take on humanity. It frames human lives as a species-wide struggle for existence in which sovereign individuals either survive or perish on the basis of their personal prowess. It presumes that human beings are born into biological destinies that no amount of socialist intervention can effectively alter. It proposes that a person's poor life circumstances result from their individual proclivities, and not from injustices perpetrated upon them that can duly be deconstructed. It implies that disparate outcomes enjoyed by social groups reflect hidden hierarchies within nature's own kingdom. Pinker's case conspicuously makes no reference to the outgroup-stigmatising discourse advanced to promote eugenics since its modern inception—not least by successive generations of psychologists.

Literally nothing in eugenics makes sense in the absence of classism, racism, xenophobia, colonialism, ableism, sexism, homophobia, or the demonisation of human frailty, totemic ills that for more than a century the liberal left has classically sought to oppose. The influence of eugenics on psychology—both historical and ongoing—starkly contradicts conservative claims that the field has been lumbered with an inherent 'liberal' bias.

12

Psychology's Conservative Paradigms

It is easy to mine historical documents for vintage quotes revealing bigotries of times past. That so many historical psychologists were socially prejudiced ought not, on its own, be that much of a surprise. Nor should it confer guilt-by-association upon today's generation. The history of psychology has bequeathed many profane talking-points that today's xenophobes can cite with glee, but ultimately talk is cheap—contemporary readers can simply disregard such utterances as anachronistic.

But in psychology's case, conservative bias influenced more than what the early psychologists once talked about. It also determined how they first pursued their craft—and so, by extension, how their successors do so today. While many modern psychologists aspire to be liberals, their foundational theories and standard methods were developed by scholars who had very different worldviews. Psychology's pioneers shaped the field's standard approaches expecting the world to submit to a conservative analysis.

Psychology's paradigms were tailored to see the world through a conservative lens: where human virtues and vices can be isolated and quantified; where welfare is rooted in personal traits and choices;

where ingroups are superior to outgroups; and where useful insights can be gleaned from traditional prejudices. The field's standards of evidence were developed to normalise the excavation of ambiguous conclusions from gerrymandered data, facilitating field-wide confirmation bias. And its professional norms evolved to encourage self-exceptionalism—acculturating psychologists to speak highly of their field and its research, to emphasise its merits and to overlook its flaws, and to defend psychology's honour whenever its outputs are criticised.

Overall, psychology remains afflicted. Its historical biases cannot yet be consigned to the pages of history books. Despite the current chorus of complaints about psychology's 'liberal' skew, today's psychologists are lumbered with an inherently conservative *modus operandi*. Psychology relies on types of reductionism that, when applied to questions of political or social justice, are intrinsically geared to favour conservative conclusions.

Conservatism by Numbers

One way to approach this issue is to ask about psychology's everyday practices. For example, what does it mean when psychologists quantify human attributes or behaviours? Psychologists are so accustomed to quantifying psychological variables that many see it as a form of logic in its own right. It is assumed that everything that humans do, feel, perceive, or even think, can reasonably be represented using numbers, with the further implication that it then makes sense to submit these numbers to mathematical analysis. Herein lies the problem.

Every schoolchild knows that if you start with two apples and somebody gives you two more apples, you end up with twice the number of apples. This is how numbers work. However, in psychology, measurement is rarely as simple as counting apples. The numbers used in psychology research hardly ever refer to actual quantities at all. Often they relate to agreement ratings, value judgements, or self-reported perceptions. What it means to quantify such entities is far from clear.

For example, a person might complete a happiness questionnaire that shows their score for happiness is 7. But what does that '7' actually mean? Let's imagine that something amazing happens in their life, and when they take the questionnaire again their new score for happiness is 14. We can certainly say that they are now *happier*, but can we say that they are 'twice as happy' as before? Is there even such a *thing* as 'twice as happy'? Does happiness really exist in quantities that can be multiplied using arithmetic? Could one person be three times as happy as someone else? Or four times? Or five-and-a-half times? For that matter, is there an upper limit on happiness or can the numbers just keep rising to infinity? In other words, *what do all the numbers actually mean*?

Aside from the philosophical challenges inherent in imagining the 'size' of someone's happiness, there are technical difficulties too. In psychology research, numbers are normally used so that data can be analysed with statistical tests. However, statistical tests are designed for very specific types of numbers and are only meaningful if certain conditions are met.

Statistical tests are usually designed for hard numbers that represent real quantities—such as those used when counting apples, where the relative sizes of numbers reflect exact differences in how many apples there are. They are not designed for numbers that are used metaphorically to describe things like happiness, where whether one value can literally ever be 'double' another is up for debate. In technical language, we say that statistical tests require variables to conform to *interval* or *ratio* scales of measurement—where the gap between each number is exactly the same, and where the number 'zero' represents the complete absence of the thing being measured (for example, a score of '0' for happiness should mean the person is experiencing no happiness at all). However, in reality, most psychological variables are just far vaguer than that. As such, in most cases they fail utterly to meet the specific scaling requirements.

Proficient psychology researchers already know this. But they have a conventional counterargument to mount in response. They say that the statistical tests used in psychology are 'robust', meaning that the damage done by violating their requirements is seldom fatal. This is a bit like saying that the ejector seat will 'probably' work. It is not entirely reassuring. The bottom line is that, in psychology, most statistical analyses are conducted in defiance of their technical prerequisites. Even when correlation coefficients, *t*-test outputs, *p*-values, and so on, are rendered to three, if not four, decimal places, the sense of precision is misleading. If the analyses fail to comply with the statistical test's terms and conditions, then, technically, all bets are off. Whether the numbers churned out by statistical software represent reliable information is down to the researcher's judgement.

Psychology's conventions of quantification raise questions about the standing of its research. Because it is widely assumed that numbers make things more scientific, psychologists often use quantification in a brain-dead way, without reflecting on its limitations (Tafreshi et al., 2016). Rarely do they think about the abstractions involved. Ratings of mental states elicited by Likert scales (where positions ranging from 'Strongly Agree' to 'Strongly Disagree' are converted into numbers) are treated as though they represent objectively verifiable quantities. Multiple such measures are combined into averages, sometimes computed to several decimal places, or else summed as if an ever-accumulating pile of apples is being assembled. The overall outcome is an implication of precision that creates an illusion of accuracy, the phantasmal nature of which is seldom addressed in the Discussion sections of research papers. This is not to suggest that quantitative measures have no place in psychology. The point is simply that it is irrational for psychologists to rely on quantitative measures without ever questioning their merits. And yet that is what most psychologists do.

Unreflective quantification creates space for bias and projection. A significant problem is the risk of scholarly hallucination, the habit of presuming that whatever is being measured *must actually exist*. But just because something has been quantified doesn't mean it is actually there. When psychologists administer separate questionnaires on 'positive emotions' and 'life satisfaction', it is usually presumed that what is being recorded—nay, *measured*—is a person's positive emotions on the one hand and their life satisfaction on the other. But this might not be the case at all. The constructs might not exist separately. They might overlap, akin to two sides of the one coin—a person's rating of life satisfaction might simply reflect the positivity of their emotions, while their rating of positive emotions might just be a byproduct of their overall satisfaction with life. Simply because psychologists can use two separate questionnaires to generate two separate scores does not mean that they are analysing two separate things (Hughes, 2016).

After a century and a half of formal research, psychology has accumulated an ocean of quantified constructs, many of which might not even exist. Most were conceived during periods when psychological discourse was dominated by social conservatism, and so are likely to reflect socially conservative assumptions. When psychologists evaluate intimate partner relationships, they tend to track outcomes that reflect traditional religious teaching—such as monogamy or commitment to parenting—even though not all happy relationships are characterised by such things. Likewise, research into group behaviour will often default to notions such as diffusion of responsibility and cognitive detachment, echoing classist concerns about the mentality of mobs. And when psychologists discuss intelligence, they frequently assume that such a thing can be reified in a unitary way and then measured with accuracy, a highly questionable claim but nonetheless a fundamental working premise of eugenics.

Qualitative methods (i.e., non-quantitative ones) have proliferated in psychology over recent decades and increasingly enjoy institutional support. However, notwithstanding such growth, the vast bulk of research in the field remains of the quantitative kind. The very habit of quantification itself rests on a belief that psychologists can divine the relevant components of human experience from numerical data. The fact that quantification is used so freely despite its many philosophical and technical pitfalls permits the persistence of stereotyped reasoning and traditionalist tropes. Far from thwarting self-fulfilling prophecies and confirmation bias, unthinking quantification serves to maintain, if not indeed to reward, such styles of thinking.

Reductionist Theories as Conservatism

Sometimes the best way to answer a question in psychology is to ignore psychology altogether. Consider, for example, the question of what might lead a person to become clinically depressed. Psychologists will usually point to genetic predispositions and neurochemical imbalances, as well as traumatic experiences, self-defeating thought patterns, and relationship difficulties. But in reality, the most reliable predictor of depression onset is not a psychological variable at all. It is economic poverty.

Poor people are up to three times more likely to be diagnosed with depression, or a related mental health disorder, when compared to their rich neighbours (Lund et al., 2010). Low household income, relative poverty, and unemployment are the strongest statistical predictors of major depressive disorder (Amiri, 2022), of mood disorders in general (Sareen et al., 2011), and even of suicide (Iemmi et al., 2016). Sudden economic downturns, such as the Great Recession that followed the global financial crisis of 2008, are invariably followed by sharp increases in community suicide rates (Barr et al., 2012). People who struggle to pay their bills are more likely to become depressed; living in a house with an average temperature less than 15 °C doubles a person's risk of mental health problems compared to living in a house where the temperature is 21 °C or warmer (Green & Gilbertson, 2008). Poverty is also associated with poor physical health (Scott et al., 2016), which in turn increases the risk of major depression (Kwapong et al., 2023). In short, economic poverty is a much more reliable indicator of depression than dopamine, serotonin, or any neurochemical. If psychologists really wanted to assess whether a person is at risk of depression, checking their bank balance would be far more useful than conducting any biochemical test.

Depression creates extensive human suffering. One in every twenty-five human beings experience it at any one time. Depressive disorders are the third most common cause of disability in the human population, accounting for more incapacitation than diabetes (GBD, 2018). It would appear that an important way to reduce all this suffering would be to eliminate poverty by promoting economic equality—in other words, by working to ensure that financial wealth is redistributed more justly across society.

However, when psychologists talk about depression, rarely do they refer to the role of capitalism or the taxation system. Instead they focus on biological causes (such as dopamine) or on other psychological specificities (such as cognitive style, self-esteem, or perceived social support). When it comes to practical intervention, they recommend remedies that focus primarily on the individual, a suite of solutions spanning talk therapy, mindfulness training, 'resilience'-building, and pharmaceuticals. Insofar as economic concepts enter the narrative at all, depression is more likely to be referred to as a 'disease of affluence' (Ridley et al., 2020), a woefully erroneous trope that entirely subverts the empirical realities of mental health.

Reductionism is the practice of analysing complex concepts by breaking them into pieces and examining each fragment in isolation. Many sciences engage in reductionism without too much controversy. Physicists often focus on particular bits of the physical world, confining their analysis within such specialist domains as particle physics, molecular physics, classical mechanics, quantum mechanics, astrophysics, thermodynamics, optics, or acoustics. Each area of physics is linked to all the others to some extent. But physicists who study photons engage in a very different type of research compared to those who study physical oceanography. The point is that all physicists reduce the physical world into discrete constituent phenomena and proceed to focus their attention on just some of them. Their efforts have resulted in several major discoveries that have fundamentally altered life on our planet.

Human psychology is also a vastly complex area of subject matter. It is no surprise therefore that psychologists too have resorted to reductionism to structure their research endeavours. For example, many psychologists reduce the human experience down to behavioural components, focusing primarily on the specific actions that make up a person's 'behavioural repertoire', while largely ignoring their cognitions, perceptions, relationships, attitudes, hormone levels, religious beliefs, or genes. Other psychologists prefer to reduce things to their psychophysiological aspects, building theories of psychology around studies of neurochemistry, electrodermal activity, brain activity, cardiovascular function, immune system responses, or even eye-movements. Reductionism per se is not inherently problematic. What makes it become problematic is when a single specialist approach is given undue weight in comparison to alternative worldviews, where broader contexts are deemed to be irrelevant, and where a particular set of reduced components is seen as sufficient to explain all that ever needs to be explained. Reductionism becomes a problem when specialists consider *their* niche approach to be the primary legitimate means by which useful insights can be produced.

For example, reductionism becomes an issue when researchers insist that psychology can be understood only in terms of biology. Biologically oriented psychologists might argue that depression cannot be properly explained (or treated) without first understanding how dopamine becomes dysregulated within the brain. Or they might claim that theories about aggression will fall short unless they account for testosterone. Or they might accuse stress researchers who don't measure cortisol of missing the point. By extension, they might also assert that brain imaging studies are intrinsically superior to all other research in psychology, simply by dint of being brain imaging studies. Other forms of biological reductionism in psychology include the belief that human behaviour is fundamentally determined by genes, or that it can be best explained in terms of evolutionary theory. Another is the hyperfocus on biological sex as the reason why women and men are said to differ psychologically (a supposition we will challenge in Part V).

The problem with reductionism in psychology is that it fails to anticipate the impact of wider influences. For example, non-biological influences might be more important than biological ones. Some symptoms of depression might well be linked to deficiencies of dopamine, but research shows that dopamine is inconsistently associated with depression overall and that neuro-chemical testing cannot be used to identify whether a person has depressive symptoms, or even whether they have major depressive disorder (Savitz et al., 2013). In other words, the majority of people with depression are physiologically indistinguishable from people who do *not* have depression. By contrast, depressed people will invariably refer to specific aspects of their life experience that they find challenging or upsetting, while major depressive disorder will frequently arise from significantly adverse circumstances or from traumatic experiences involving abuse, interpersonal conflict, or exploitation. And, of course, economic conditions will be one of the strongest predictors of all.

The idea that depression is best addressed using psychotherapy or drugs rather than by wealth redistribution is just one example of how reductionism affects the politics of psychology. Similar skews arise when psychologists consider other social problems. Violence and aggression are normally discussed in terms of (masculine) impulsivity or testosterone levels, as opposed to cultural toxicity, societal neglect, or economic disempowerment. Chronic feelings of mental stress are framed as arising from sustained physiological arousal—something that can be addressed by breathing exercises or cognitive reframing—with few psychologists focusing on the injustices or exploitation that stimulate this arousal in the first place.

Even political turmoil is cast as something that personal remedies can best deal with. When experts on BBC Radio 4's *PM* show first advised listeners on how to deal with the outcome of the UK's Brexit referendum, they suggested that the best reaction to political disappointment would be to 'take control of the things [you] can control, such as sleeping, eating, exercising, and limiting [your] exposure to social media' (Degerman, 2019a). It was as though any form of *political* response—such as taking to the streets in protest or participating in an organised campaign of opposition—would, in mental health terms, be ill-advised. Indeed, psychology's narrative on social anxiety frequently appears to frame concerted political dissent as, in itself, something pathological. We will return to this problem in Part V.

Political bias is inherent in psychology's reductionism, an explanatory tic that places the ultimate locus of all causality within individuals, while downplaying broader social, economic, and political contexts that might in fact be pivotal. Psychology's reductionist narratives suggest that human predicaments are the outcomes of individual experiences and attributes, thus implying that they are best addressed by solutions that share this assumption.

Reductionist Methods as Conservatism

Reductionism is so intrinsic that it is reinforced by virtually all of psychology's standard research methods. Correlation, for example, is one of the classic tools of reductionism. Not only does it hark back to nineteenth-century eugenics, correlation also perpetuates a clichéd form of reasoning that elides nuance, seeks meaning from coincidence, and imputes relevance from juxtaposition. Virtually all undergraduate students are aware that *correlation does not imply causation*—but the problems with correlation extend much further than the spurious attribution of cause-and-effect.

The key problems relate to concepts rather than numbers. Because of their simplistic structure, correlations create the impression that highly complex situations can be explained using a relatively small number of ideas. But psychological outcomes are nearly always shaped by hundreds if not thousands of inputs, far more than are covered by the two included in a bivariate correlation (or the ten or so that might be incorporated into a relatively packed regression model). The formulaic way in which correlations are conceptualised will always fall short of reality. Nonetheless, the very act of analysing data in this manner serves to reify correlations in psychologists' minds. It drives their conclusions toward simplistic explanations. In life, the wisdom of experience usually teaches us that events are more complex than they first appear; concepts such as correlation serve to *impede* the development of such forms of wisdom in psychology.

Something similar can be said about methods that focus on group differences rather than correlations. Many studies in psychology compare the scores of different groups of people, frequently condensing the varying attributes of hundreds (if not thousands) of people down to one or two data points. The implied binomial nature of a two-group comparison belies the expansive psychological complexity that will always be inherent within any sizeable group of human beings. It takes only a moment's reflection to realise that a large group of women, for example, will contain a broad range of human characters, each with unique experiences, outlooks, traits, and capacities. And yet, in any study in which women are compared to men—as psychologists frequently do when examining supposed sex differences in mathematical ability, leadership skills, visuospatial reasoning, or self-esteem—all these women will be combined and considered as a unified block of information, from which a single representative score can be computed and contrasted with an equally reductionist single-number score drawn from a just-as-diverse group of men.

The reductionism inherent in group-comparison research is bad enough when those groups exist in the natural world. However, in reality, most groups that *appear* to exist in nature do not exist as simply as we often presume. Human sex itself is an example. Biological sex in humans is often discussed in terms of a simple two-group dichotomy of females and males, when in fact around 1.7% of the human population (about the same number as have ginger hair) defy the gender binary by exhibiting nondimorphic sexual development (Blackless et al., 2000). This means that, in addition to condensing massive group complexity into paltry single scores, the very fact that psychologists use two-group designs at all to study sex differences is *itself* an example of reductionism.

The problem of group reductionism is even more pronounced when psychologists attempt to create groups from what are otherwise ungrouped datasets, such as when they use arbitrary cut-off scores on personality questionnaires to distinguish 'introverts' from 'extraverts'. Most psychology students will be familiar with the warning that arbitrary dichotomies should not be treated as though they were actual groups. However, cut-off-score reasoning is widely employed in clinical psychology, for example, when psychometric tools are used to screen patients for particular mental health diagnoses.

Reductionism also underpins the use of technology in psychological research. An archetypal example of this is brain imaging. While the machines concerned are truly breathtaking, the research designs of the studies that employ them are rarely that sophisticated (Hughes, 2018a), and their sample sizes are almost uniformly underpowered (Button et al., 2013) and unrepresentative (Chiao, 2009). Despite this, brain imaging research is frequently touted as epitomising psychology's cutting-edge approach to science.

The belief that psychology should project itself as a technology-based research discipline dates back to its formative years in universities, when Wilhelm Wundt stocked his Leipzig laboratory with tuning forks and stopwatches. In the words of critical psychologist Isaac Prilleltensky, psychologists have learned to seek 'salvation through science and technology' (1994, p. 19).

However, the rise of the machines produces a number of undesirable outcomes. Firstly, it bolsters the reductionist idea that looking inside a person—in this case literally inside their head—is the best way to find out how they are doing psychologically. It thereby encourages us further to ignore the impacts of social inequalities, political policies, or cultural forces. Secondly, by implying that psychology's research processes can be automated, it avoids drawing attention to potential distortions that might result from the investigator's own values, morals, or ethics. And thirdly, it promotes an instrumentalist approach to problem-solving, where material resources (which cost money) are heralded ahead of intellectual, collaborative, or democratic effort (which should be free). Such technophilic privileging reflects a broader neoliberal value system, in which the worth of individuals, organisations, and even countries is to be judged on the basis of their accumulated capital and wealth.

Few matters in psychology are easily describable in simple terms. Considering events only in terms of narrow fragments of interest, such as biological precursors or cognitive perceptions, necessitates the ignoring of broader, holistic truths. Reductionism ensures that psychology focuses on oversimplified versions of reality. As detailed as psychology research can sometimes appear, real life is typically far more complicated. Even the most elaborate psychological study will confine itself, at best, to a dozen major variables (researchers who attempt to look at more will be criticised for over-stretching), whereas a single human action will be influenced by thousands of determining factors, many of them intangible and immeasurable. In this way, psychologists are left with methods for explaining the human condition that cohere to the dynamics of conservatism—complex, abstract, and potentially challenging insights are resisted in favour of hasty stereotypes, coarse generalisations, and self-serving summaries of other people's problems as products of their own creation.

Individuocentrism as Conservatism

Reductionism is a legacy of psychology's history. It is how things are done because it is how things have been done in the past. In a sense, the entire field of psychology could be said to have emerged from the habit of reductionism. Had it developed differently, psychology's remit could reasonably have spanned social systems, environmental conditions, historicopolitical discourse, communal knowledge, and cultural change. But psychology did not evolve that way. Instead, it reduced the task of studying human experience to a science focused purely on the perspective of individuals.

This is true even of social psychology, that part of psychology that is supposed to consider human interactions and relationships. While some social psychologists study communities, societies, and other pluralised entities, a predominant majority now confine themselves to looking at 'the individual in social contexts'—focusing not on group-level consensual understandings, but on how *individuals within a group* come to perceive and react to other people (Allport, 1954). This work tends to comprise laboratory studies in which participants are asked to complete tasks in the presence of others (a classic example being Milgram's famous obedience experiments), or else large-scale questionnaire surveys probing people's 'social cognitions', from which statistical models purporting to explain their susceptibility to social influence might be derived. Over the course of the twentieth century, social psychology refined itself into a largely individuocentric pursuit. Consideration of the behaviour of groups (including that of 'mobs') gradually gave way to an almost exclusive focus on the behaviour of the individuals within those groups.

Indeed the dominance of this individualised form of social psychology is such that the more 'social' version is now customarily ostracised by psychology's gatekeepers. So much so that Wikipedia currently hosts two separate pages under the title 'Social psychology'. The first describes the *individuals-in-social-context* approach to the field, positioning it as a bona fide subfield of psychology. The second page, which describes a century's worth of scholarship looking at societies, cultures, identities, and groups, is pointedly titled 'Social psychology (sociology)'. It baldly insists that this type of thing is a 'subfield of sociology' despite the fact that most of the work it describes was produced by psychologists working in university psychology departments. (For example, all three editors-in-chief of the *European Journal of Social Psychology* work in psychology departments, while the similarly scoped *British Journal of Social Psychology* is edited by such a psychologist and, for good measure, is published by the British Psychological Society; Hughes, 2023).

Some observers argue that the displacement of sociocentric perspectives by hegemonic individuocentrism amounts to the 'Americanisation' of their field (Marková, 2012). It certainly reflects a broader transatlantic divergence across many areas of academia where US norms have come to marginalise their European counterparts in similar ways. The 'individualisation of the social'—the focusing of social psychology on individual experiences—first appeared and thrived in North America, whereas the tradition of considering socially shared experiences was initially more prevalent in Europe. As we will discuss in Part VI, over the past century, American academia has become increasingly culturally dominant. Psychology's overall research agenda is now driven by, and thus reflects, American interests.

As a consequence, psychology has been shaped by the gradual marginalisation of what could otherwise be regarded as entirely legitimate scholarly approaches. In the 1960s, precocious intellectuals such as Jean-Paul Sartre attained Nobel-worthy renown for their existentialist theories of human agency and decision-making, and yet their ideas were deemed insufficient for scientific psychology, where (American) behaviourism taught that all human actions were in fact subject to external influence and control. That Sartre himself discussed behaviourism as an 'obsolete position' was to prove ultimately of no avail (Amouroux & Zaslawski, 2019). During the same period, Soviet-world theories of human nature—such as dialectical materialism, with its premise that human beings were instinctively primed to accumulate property at the expense of their neighbours' welfare—were effectively boycotted by Western psychology. The Russian psychologist Alexander Luria is today typically described in Western textbooks as having been little more than a neuroscientist, while his social theories of 'cultural historical psychology' (as well as his harsh criticisms of behaviourism) are air-brushed from psychology's official histories.

And in the twenty-first century, the vast bulk of psychology's research activities are now focused on a tiny subset of the human population, namely people living in so-called WEIRD societies—places that are Western, Educated, Industrialised, Rich, and Democratic. The field effectively accumulates its evidence by studying a small sliver of the human population who live in largely individualist cultures, rather than from majority-world populations who reside in places where social collectivism is prioritised as a cultural value. We will return to the problems created by this geographic skew in Part VI.

In short, today's psychology is a field that examines individualised experiences from the point of view of individualist cultures. Its historical development as a science has gradually—but effortfully—discarded any intellectual influences that might emphasise the role of society or culture on human behaviour. Such approaches have been delegitimised as being insufficiently 'psychological' and thus, by implication, 'unscientific'. Psychology, we are often told, simply *is* the science of individuals. Its focus on the singularity of personhood—rather than on, say, the commonalities of groups, of societies, or of cultures—is acclaimed as its unique selling point, a feature rather than a bug, the element that elevates it beyond all the other behavioural sciences. As 'value-free' and 'objective' researchers, psychologists are minded to leave society to the sociologists, economics to the economists, and politics to the political scientists (or perhaps even to politicians).

In attaching primacy to individual experiences, emphasis is removed from external forces. Social 'realities' are thus reified. Plights such as poverty or criminality are seen as person-level phenomena, rather than as products of unjust economic systems or uncurtailed avarice. Social problems are framed as personal challenges. When psychologists talk about global pandemics, the climate catastrophe, or other existential threats to the planet, their narratives enforce a focus on individual action, rather than promoting social policy interventions that might protect the vital ecology upon which all human health, including mental health, is dependent. Psychology's individuocentrism is entirely consonant with the 'personal responsibility' narratives so favoured in conservative neoliberal politics.

Locating human agency solely within an individual's personhood avoids the essential fact that people's lives are intertwined in complex social, economic, political, historical, environmental, and interpersonal realities. By contrast, psychology's driving narrative is that people can and should act alone, that the existence and interests of other people are divorced from your life realities, and that other people have little or no responsibility to address what are *your* problems.

In short, psychology's individuocentrism implicitly serves to absolve human beings of the obligation to consider externalities when deciding their own actions, permitting them to disregard the impacts of their choices on other humans, on animals, or on the physical environment. In this respect, psychology and capitalism share a single political outlook—namely, a belief in the centrality of the 'supreme self' (Prilleltensky, 1994). Psychology's individuocentrism is thus inherently conservative. It superficially reinforces neoliberal narratives of personal responsibility, valorising the so-called 'sovereign individual' who possesses ultimate individual rights.

13

From Conservative Past to Neoliberal Present

The juggernaut of history has endowed psychology with inescapable legacies. It has moulded it into a field that trades in unreflective quantification, unwarranted group comparisons, and unjustified causal stereotypes. It has bequeathed to psychology a set of methodologies, some of which were devised to perpetuate conservative agendas, others of which were simply shaped by conservative thinking. It has led psychologists to probe group-level differences (while ignoring the diversity of the groups themselves); to focus on rudimentary correlations and regressions (while disregarding the hypercomplexity of human life); to look for signals within the noise (while failing to ask, what does the noise mean?); to imagine that they have all the information they need within whatever statistical dataset they have assembled; and to embrace a habit of reductionist reasoning that locates psychological cause-and-effect within people as we superficially see them, rather than within the opaque and sometimes invisible realities of ecological, societal, and political contexts.

There is a stale old cliché to the effect that if the only tool you have is a hammer, then all your problems start to look like nails. Psychology's many hammers were designed, over a century, to divide humanity into hierarchically arranged groups, to validate the sanctimonious norms of

© The Author(s), under exclusive license to Springer Nature Switzerland AG 2025   **109**
B. M. Hughes, *Psychology's Quiet Conservatism*,
https://doi.org/10.1007/978-3-032-07724-0_13

Victorian traditionalism, and to construe life itself as an exercise in personal sovereignty, in which there is little or no need to blame society for society's own ills. While today's psychologists are certainly often accused of displaying a liberal bias, their training, theories, and go-to methods have evolved in ways that acculturate them to *think like conservatives*.

The consequence of this history has been to enmesh psychology's accounts of human welfare and improvement within the fallacious circular logic of individuocentrism. In the next few chapters we will examine the key problem of victim-blaming in psychology—how cardinal concepts compound the conservative narrative that people are responsible for their own life circumstances.

In the end, the structuring of human welfare as a personal responsibility of individuals is politically non-neutral because it falsely assumes that power is distributed equally across society. The framing of 'positive thinking' as integral to human well-being—and the associated stigmatising of pessimism as a 'self-destructive' thought style—combine to suppress political dissent. By teaching people that they should face up to, and accept, their own powerlessness in the face of (political) events, psychology explicitly encourages humanity at large to find satisfaction in the status quo, rather than to challenge it.

As all human societies are in fact characterised by socioeconomic inequality, the standard advice from psychologists, by leaving existing power hierarchies unthreatened, actively serves to perpetuate societal injustice. Psychological maxims serve as an opium of the people, deployed to calm popular unease about the nature of society. Psychology's notions of human well-being have come to be framed in ways that protect the heritage of domineering traditionalism that shaped the field for more than a century. They reflect a worldview that is laden with historical conservatism, and which is very far from liberal.

Part III

Psychology, Capitalism, and Human Welfare

14

Hierarchies and Hysteria

The concept of 'mass hysteria' is one of the most widely discussed psychological notions in contemporary culture, while equally being one of the most chauvinistic. It combines into a double-diatribe both the threat of 'the masses'—those dangerous plebeians so feared by the upper classes—and that of 'hysteria'—a sexist epithet derived from the ancient belief that a wandering uterus engenders (in women) ungovernable emotional excess. As discussed in Part II, psychological discourse has long posited the existence of a societal underclass while simultaneously dismissing difference as pathology. Eighteenth-century scholars decried as deranged the voteless mobs who sought, through violence, to overthrow kings and institute democracies. Their twenty-first century counterparts have been just as vitriolic in their condemnation of so-called woke social-justice warriors who deign to believe that climate justice might be a human right, that refugees should be welcome, that trans women are in fact women, or that the lives of non-White people might actually *matter*.

It is in the nature of hierarchies that power is often embedded in the everyday in ways that make it less stark than fables might have you believe. Not all establishment figures wear crowns or live in big

© The Author(s), under exclusive license to Springer Nature Switzerland AG 2025   **113**
B. M. Hughes, *Psychology's Quiet Conservatism*,
https://doi.org/10.1007/978-3-032-07724-0_14

houses. Privilege is not always seeped in pomp and ceremony; quite often it hides in plain sight. It is structured across many levels. At the top of things are the oligarchs and billionaires, but there are many layers between them and the meek and hungry. Power percolates downward such that most people are beholden to some but simultaneously privileged over others. A classic old comedy sketch written by comedian Marty Feldman and writer John Law once showed three men discussing their relative statuses (Dewsbury, 2016): 'I look up to him because he is upper-class, but I look down on *him* because he is lower-class,' one of them declares. 'I,' he then explains, 'am middle-class.'

This bidirectionality of perspective helps to highlight how everyone enjoys privilege of some sort. Everyone can look down on someone. Those skilled at preserving their own high status tend to be good at finding ways of ensuring that ordinary people—those in the middle—are always looking down on others in their midst, rather than gazing upwards at society's true power brokers. Dismissing dissent as mass hysteria is a common example of how controversies can be blamed not on those who create injustices, but on those who raise grievances. Depicting complainants as dangerously irrational causes us to pity them, to look down on them, rather than to consider the possible merits of their case. It is perhaps no wonder then that so much political energy is spent on scapegoating.

Gatekeeping Academia

The practice of self-preservation through blame-misdirection is not exclusive to politics. It can be encountered in any hierarchical system, whether global, national, community-based, local, or organisational. And few societal systems are more hierarchical than academia. Universities tend to pride themselves on their traditions, but most of those traditions were shaped in more feudal times. Academia is an occupation grounded in ranks.

While the nomenclature varies by university, usually the professional ladder begins at something like 'teaching assistant' and ends, once sufficient dues are paid, at 'professor'. In many universities even the rarefied professors are drawn into the comedy of looking up and down at each other. Apart from the honorary professors, adjunct professors, and emeritus (i.e., retired) professors, full-time profs will likely be stratified into several nuanced ranks-within-ranks: assistant professors, associate professors, full professors, endowed professors, titular professors, distinguished professors, and the like. (My own idiosyncratic institution adds 'personal professors' and 'established professors' to this pile; ostensibly the same but different, just with some more equal than others.)

The true standing of academic titles will be different in different places. In the United States a 'professor' can be quite a junior drone, whereas 'professor' titles in the United Kingdom are reserved for senior bods only and command considerable deference (not to mention remuneration). A *professeur extraordinaire*, meanwhile, can be an esteemed guest scientist who visits your university (in Switzerland) or a part-time academic (in Belgium). Or, I suppose, just a professor who is extraordinary.

The Best Form of Defence

Academic status anxiety frequently leads to territorialism. As a result, many academics are poorly disposed toward the prospect of non-academics challenging their ideas. In the spring of 2021, British journalist George Monbiot wrote a high-profile article in *The Guardian* warning readers about long COVID, at that time an emerging but still poorly understood public health threat. He described how COVID-19 infection had for many sufferers produced long-lasting impacts on major systems of the body. Moreover, he discussed how these effects closely resembled the symptoms of myalgic encephalomyelitis or chronic fatigue syndrome (ME/CFS) and explained how ME/CFS had been 'disgracefully neglected by science and medicine' both in the UK and around the world. He argued that the British health service should 'learn from its mistakes' and start taking long COVID more seriously (Monbiot, 2021a). The article made an immediate splash and helped introduce the topic of long COVID to mainstream audiences in the UK.

However, Monbiot was quickly condemned by a senior psychology professor who accused the journalist of not knowing what he was talking about. According to the professor—who worked at a prestigious university and held several influential roles in Britain's professional bodies— Monbiot had missed the point of long COVID by failing to understand that it was, in fact, a *psychologically driven* condition. Indeed, the professor felt that simply by mentioning long COVID in his journalism, Monbiot was guilty of endangering the public. Given that the illness was psychologically driven, he suggested, press coverage like Monbiot's would be a key factor in stimulating its spread—reading about the condition would simply cause more people to turn up to clinics complaining of long COVID symptoms, symptoms caused not by physiological damage resulting from coronavirus infection but by psychological dysfunction arising from excessive health anxiety. As Monbiot himself put it, the professor was accusing the journalist, directly, of being a one-person long-COVID 'super-spreader' (Monbiot, 2021b).

The narrative was clear. According to the professor, people like Monbiot should know their place. They should stay in their lane. They should leave psychology to the psychologists. Monbiot was just a writer, and a reckless one at that, whose inability to understand the psychological complexities threatened to do more damage than good.

The professor viewed long COVID through the lens of mass hysteria. Accordingly, he believed that the best way to treat long COVID would be to use 'psychologically informed rehabilitation' (Monbiot, 2021b). It was perhaps no coincidence that this very same professor had for years recommended 'psychologically informed rehabilitation' as a treatment for many conditions, including ME/CFS. He had published several research papers claiming that post-viral syndromes could successfully be treated using approaches based on cognitive behaviour therapy. The professor was not alone in having invested his reputation in this way. He was one of a number of high-profile academic psychologists who claimed that post-viral conditions such as ME/CFS (and now long COVID) were essentially 'psychogenic' illnesses—in other words, illnesses rooted in psychological processes and so best treated using psychological therapy—despite the fact that an increasing body of medical research had exposed that notion as spurious.

Treating a medical condition using psychotherapy is typically cheaper than treating it using drugs or surgery, and people who only *think* they are sick are usually not entitled to sickness benefits. This means that psychogenic explanations of conditions such as long COVID have important financial implications for the state. Simply put, medical protocols that classify a person as psychologically unwell, rather than physically sick, help to save money. Delivering psychotherapy to outpatients is cost-effective. No hospital beds are needed, no expensive drugs or surgeries are involved, and professional psychotherapists have lower salaries and are cheaper to train than medical doctors. Setting up a system of 'fatigue' clinics that rosters patients onto weekly psychotherapy sessions is exceptionally good value, especially if there is no regulatory pressure to prove that patients are actually getting better. Overall, 'psychologising' these conditions is highly attractive to the fiscally parsimonious. It reduces demand on public healthcare budgets—thereby helping government to keep taxes low—and boosts the profits of private healthcare providers and medical insurance companies.

In addition, people who have ME/CFS or long COVID are typically unable to hold down jobs, making them useless to the capitalist economy. Instead they end up relying heavily on their families and the wider community. Even then most sufferers descend toward impoverishment. All of this presents a challenge to conservative politics because it shines a light on the importance of wealth redistribution, social safety nets, disability rights, investment in public services, and other forms of government interventionism. As a consequence, people who suffer from conditions like long COVID or ME/CFS will frequently find themselves targeted by right-wing opprobrium. Psychological concepts like malingering, health anxiety, illness perceptions, and mass hysteria provide highly convenient attack lines for those wishing to discredit the demands of the disabled. It is disturbing just how frequently so many esteemed psychologists—purportedly 'liberal' academics at that—end up actively enabling these attacks.

15

The Contrivance of Capitalist Minds

'If you think you're free,' the American comedian Bill Hicks once said, 'try going somewhere without money' (Jackson, 2014). Hicks was not alone in identifying (and questioning) how humanity has allowed money to become central to its existence, perhaps even more central than 'freedom' itself. Money is a resource that has become more or less necessary for a person to survive. This puts it on a similar level to water, food, and air. The difference is that water, food, and air are naturally occurring entities that cohere directly with the physiological processes that constitute organic vitality. The very idea of human survival is synonymous with biological sustainability—the fact that water, food, and air are essential to life is another way of saying that life itself constitutes the chemical incorporation of these biological entities into healthy bodies. Water, food, and air are physiological fuels without which bodies cannot function. This is a literal fact, not a metaphor.

However, money is different. Money is a social convention. It is an arbitrary system that human beings invented and have agreed, at least implicitly, to operate in perpetuity. It is a way of facilitating trade by recording exchanges of assets based on the socially perceived value of those assets. In times gone by, money used to be linked to specific

© The Author(s), under exclusive license to Springer Nature Switzerland AG 2025
B. M. Hughes, *Psychology's Quiet Conservatism*,
https://doi.org/10.1007/978-3-032-07724-0_15

possessions, such as gold or other precious metals. But nowadays money is usually just symbolic. Currency is issued by governments and declared, by fiat, to have immediate value—to be 'legal tender'—as part of a wider agreement between governments and banks to regularise monetary exchange. Money can also be demanded in the form of rent, fines, interest, or access fees, without there being a clearly quantifiable cost borne by the provider—essentially money for (literally) nothing. Even money that is backed by a particular commodity, such as a stated quantity of gold, will usually rely on social convention to determine its value. After all, gold might be shiny and fashionable, but it is of limited practical use to most people. Its value derives from what people *believe* its value should be. In other words, its value is determined by how much people are willing to pay for it. Money, then, is largely a circular concept.

The idea that money is essential to human existence only makes sense if we presume that human beings *by nature* require a system for quantifying and exchanging their possessions. Notably, insofar as we are aware, no other species requires such a system. In reality, it is very questionable whether human beings really do require money or whether the accumulation of wealth is an authentically natural human impulse. It could simply be that our appetite for wealth accumulation is a byproduct of our instinct for social conformity—we feel the need to seek out wealth because that is what we see *other* people doing. Lots of crazes emerge in culture that owe their existence to conformity behaviours. In the eighteenth century, many Europeans became suddenly interested in pineapples; not only seeking to grow the fruit in colder European climates but also incorporating images of pineapples into their decorative art, furniture design, tableware, architecture, and even hairstyles (O'Connor, 2013). So-called 'pineapple fever' persisted for a century-and-a-half, driven not by necessity but by an apparent impulse for imitation. Europeans developed a penchant for pineapple-themed wallpaper not because it was useful, but because it was considered *normal*. In a similar way, the social convention that is 'money' has become normalised. When born into a world in which everybody around us is using it, it is understandable if we come to unquestioningly believe that money is something *we* need to be using as well.

Wealth accumulation might well be a psychological hyper-elaboration of an otherwise lower-order impulse, such as the impulse to gather food or to find shelter. It is natural for human beings to seek to possess *some* things, at least those things that are necessary for our physical survival. But it is less obviously natural for us to seek out extreme wealth, such as that involved in becoming a billionaire. Strictly speaking, we can get by with three meals a day and a warm place to sleep. The exigencies of modern economics lead us to seek out far more than we need, so much so that humanity is required to operate its global system of money so that exchanges of wealth can be audited. Money, and by extension capitalism, have become ubiquitous in the modern world. But that does not mean they are truly necessary for human life to be sustained. We could, for example, *share* our resources, rather than trade them in exchange for money.

In many ways, our attitude to wealth might simply be a rational response to the circumstances in which we find ourselves, and which have arisen due to forces beyond our control. Being born into a capitalist economy means that we must engage in capitalism in order to survive. In many capitalist countries, having no money means going without healthcare, education, legal representation, and adequate nourishment. Quite literally, engaging in capitalism is required to put food on the table. The underlying human impulse here is not a need for money and capitalism, but a need to not *die*.

Reality Bites

The very fact the world economy adheres to a largely capitalist approach makes many people believe that capitalism is the only viable system for organising human affairs, a worldview that British philosopher Mark Fisher once called 'capitalist realism' (Fisher, 2009). At its core, capitalist realism is a theory of psychology. It sees capitalism as reflecting innate human desires and hypothesises that all other systems must eventually fail because of their inherent incompatibility with human nature. This take on the psychological imperative for capitalism entered political folklore in the 1980s with the slogan 'There Is No Alternative', or TINA, and was commonly associated with former UK prime minister Margaret Thatcher (as a consequence, Thatcher was occasionally nicknamed 'Tina' by her critics; Burns, 2012).

However, there *are* alternatives, both practical and theoretical. While definitions can always be debated, several countries and territories have operated economic approaches that clearly deviate from orthodox capitalism. These include places such as China, Vietnam, Laos, and Cuba, all of which describe themselves as socialist countries that aspire to institute future communist societies. Each of these states engages in international trade within the global capitalist economy and all operate currencies that their inhabitants use to acquire products. However, their internal economies are organised around nationalised state-owned enterprises in which the government acts centrally to manage the country's overall resources (or 'means of production'). This system is sometimes referred to as 'state capitalism' (e.g., Bremmer, 2010), in order to distinguish it from the traditional form of capitalism to which Thatcher felt there was no alternative. Other countries, such as a number of states in West Asia with petroleum-based economies, centralise their economic systems without dallying with communism per se, and are sometimes therefore described as being *neither* capitalist *nor* socialist. A notable feature of all these non-capitalist countries is that they tend to be one-party states (either officially or unofficially), in which democratic elections are highly circumscribed and political dissent is generally not accommodated.

In reality, for good or for ill there are several alternatives to the form of capitalism that Thatcher and others discussed. A variety of economic and political systems can be, and have been, instituted around the world. Just as there is no 'pure' form of communism, likewise there is no 'pure' form of capitalism: by far the majority of capitalist countries use taxation to subsidise public services and provide social welfare safety-nets to citizens who are unable to acquire wealth on their own. If anything, this suggests that human beings have a strong impulse to provide support to one another at a community level. The claim that 'pure' capitalism reflects innate human psychology and that any alternative system is doomed to fail because it will run counter to human nature is largely mythical. Capitalist realism is a widely held belief, but it offers a stereotype of human psychology that is demonstrably questionable.

16

Capitalist Psychology

Nonetheless, academic psychology has embraced the stereotype of capitalist realism. Psychology too appears to presume there is no alternative. The human being described by mainstream psychology is one who pursues self-interest and is incentivised by individual reward. They accumulate resources and avoid indebtedness or loss, and weigh the merit of an action by computing its profitability in terms of costs and benefits. Socially primed for capitalism, *Homo psychologicus* constructs utilitarian relationships, seeks equity in social exchange, competes for advantage against peers, and is driven to pursue a life of economic productivity. If they prove *incapable* of economic productivity (or if they are simply indifferent to the idea), psychology will likely classify them as disabled or mentally ill.

Psychology's major paradigms exemplify this capitalist realist framing. Behaviourism depicts human motivation as arising from a person's automatic response to gratification, thereby normalising gratification as the default basis for behaviour. Actions and choices are portrayed as being shaped by the rewards they generate from the environment. These rewards are then construed as the building blocks of human psychology. According to behaviourist theory, the best way to stimulate a person to

© The Author(s), under exclusive license to Springer Nature Switzerland AG 2025 **125**
B. M. Hughes, *Psychology's Quiet Conservatism*,
https://doi.org/10.1007/978-3-032-07724-0_16

behave 'well' is to offer them the right type of reinforcement—to pay the correct price, as it were, as if such things are commodities that can be bought and sold. Clinical behaviourists even go so far as to say the quiet bit out loud, recommending the use of so-called 'token economy' interventions, systematic programmes in which clients accumulate literal tokens as rewards for 'positive' behaviour that they can later exchange for tangible booty. With behaviourism, all human actions are determined by incentives. ('Try going somewhere without material reinforcement' appears to be the behaviourists' variation on Hicks.) Even generosity and self-sacrifice are interpreted in terms of how they circuitously satiate the do-gooder's own desires.

Cognitive psychology, for its part, is usually seen as offering an antithesis to the baldness of behaviourism. But it too embraces the capitalist realist view of human nature. Cognitivists classify decision-making as 'rational' only insofar as it advances the decision-maker's own self-interest. An individual whose choices operate *against* their self-interest is said to be thinking irrationally *by definition*. Cognitive theories imagine that human beings are constantly weighing up 'costs' against 'benefits', ostensibly treating every life choice like a business opportunity. Self-advancement or, at the very least, self-preservation are the benchmarks against which the quality of this process is to be evaluated. Charitable impulses, aesthetic choices, the desire for social communion, and even conscientiousness and guilt are treated as 'contaminating' factors that 'distort' human judgement, 'fallibilities' produced by people's perilous proclivity for 'fast thinking' (Kahneman, 2012). Cognition is construed as a form of calculation devoid of human compassion (Teo, 2018). Self-sacrifice and spontaneity are presented as examples of the human condition's occasional quirkiness, rather than as the very thing that defines us as being different from machines.

Even humanistic psychology, an ostensibly 'woke' subfield, is as likely to get sucked in. Humanistic psychology is typically framed as promoting awareness, thoughtful action, and empathy. Nonetheless, humanistic psychologists identify *self*-actualisation as the pinnacle of psychological functioning. Their field idealises individual fulfilment, rather than deference to others, as the goal of personal development. Many humanistic psychologists are deeply concerned about social issues, and the field has a long tradition of championing social justice causes. But the person-centred focus of humanistic psychotherapy encourages a particularly individualistic form of humanitarianism, in which a better world is to be achieved one self-actualisation at a time (Prilleltensky, 1994). The problem is that individual growth does not automatically lead to societal improvement—social inequalities are rarely resolved without wider political change and the disruption of power hierarchies. In fact, humanistic psychology's emphasis on self-actualisation, personal potential, free will, subjective experience, positive thinking, individual fulfilment, and personal growth could ultimately make social improvement harder to achieve, rather than easier. Encouraging people to focus on self-awareness and on cultivating more 'intentional' reactions, rather than teaching them *to critique and challenge the reactions of others*, draws attention away from the importance of shared responsibility and ensures that power hierarchies are left unthreatened.

Subfields of applied psychology practice have also become encumbered by the capitalist realist stereotype. For example, educational psychology largely views students (and lifelong learners) as self-contained individuals, competing for credentials with which to enhance their employability within the capitalist economy. The focus is on the role of person-level factors in educational attainment, such as self-esteem, goal setting, engagement, vocabulary, fluency, comprehension, inference skills, learning anxiety, 'mindsets', or 'learning styles'. This individuo-centric emphasis distracts from the cultural, societal, and historical biases that profoundly undermine the life prospects of so many marginalised children. Educational psychologists evaluate success in terms of how individual students perform academically, instead of in terms of how the education system might benefit society as a whole, by (for example) promoting altruism, social cohesion, or respect for the environment. Some education researchers do examine the impact of education practice on outcomes such as social stratification and cultural assimilation, but they are usually sociologists, not psychologists. Psychology's toolkit is simply not designed to hammer such communitarian nails. Instead, psychology sticks to its traditional neoliberal brief, focusing on employability rather than enlightenment as the intended end-goal of education (Sugarman, 2015).

And speaking of employment, if there is one field of applied psychology that is especially drawn toward capitalist norms then it is perhaps, obviously, occupational psychology (intermittently referred to as business, industrial, or organisational psychology). The British Psychological Society defines occupational psychology as 'the application of the science of psychology to work' (British Psychological Society, 2024a). According to the BPS, the aim of occupational psychology is 'to deliver tangible benefits by enhancing the effectiveness of organisations and developing the performance, motivation and wellbeing of people in the workplace.' But it is not immediately clear to whom these tangible benefits are supposed to be delivered. There is in fact an inherent tension between what employers and employees might regard as 'tangible benefits'. Whether in the profit-making or non-profit sectors, employers are rarely motivated to inflate their employees' wages or to improve their working conditions, especially if it pushes operating budgets into the red. No number of employee workshops on mentoring, life-planning, or leadership can hide the fact that the employer would offer none of these experiences if it hurt their bottom line over the long term. And the very fact that occupational psychologists participate in industry *at the behest of employers* requires that they take the side of the bean-counters. Ultimately, the language of occupational psychology betrays a narrow set of pro-management capitalist assumptions: it implausibly imagines an industrial landscape devoid of class conflict, where exploitation never occurs and in which executives and janitors are collegially engaged in pursuing a common goal (Prilleltensky, 1994).

Clinical psychology is the field of applied psychology that seeks to outline what is meant by psychological dysfunction (or in older language, 'abnormality') and then use that knowledge to support the wellbeing of people who experience such dysfunction. More succinctly, clinical psychology encompasses the part of psychology that defines and treats mental illness. In reality, however, psychologists have rather vague notions about what constitutes mental illness per se. As we discussed in Part II, there is no systematic way to define psychological 'abnormality' and, when it comes down to it, all such diagnoses tend to be based on arbitrary social conventions. People can be declared mentally ill simply by virtue of being *different*, meaning that social norms become the benchmark against which madness is judged. If those social norms are themselves bound up in capitalist realism, then clinical psychology itself becomes a means by which to police people's attitudes toward capitalism.

Several philosophers and historians have noted how psychology's methods of diagnosing mental illness tend to cohere with capitalist assumptions. For example, in the centuries prior to the Industrial Revolution, people with poor cognitive abilities or eccentric personalities tended to live conventional lives within the mainstream community. However, after the emergence of modern economic systems—which relied on numeracy, bureaucracy, technology, and tight legal regulation—such individuals struggled to participate in what society now required of them (Porter, 1987). This was one of the reasons why mental asylums proliferated in Europe from the seventeenth and eighteenth centuries onward. The fundamental transition led to a societal practice that the philosopher Michel Foucault called 'le grand renfermement', or the Great Confinement (Foucault, 1961). As the modern world became focused on order, efficiency, and documentation, the privileged classes set about identifying people who were 'unfit' for the new economic reality, diagnosing them as mentally ill, and forcibly confining them in their thousands to such civic institutions as asylums, poorhouses, and prisons. In the twentieth century, psychologists broadened the field's classifications of intellectual disability, drawing on eugenic theories to promote the sterilisation of what they deemed to be 'mental defectives', lest these people's presence prove too much of a drain on society's economic resources (Ilyes, 2020). The ability to be productive—in economic terms—became part of what was seen as psychological well-being. This fundamental expectation that people are capable of fitting in with their surroundings remains a cornerstone of mental health diagnosis in contemporary clinical psychology. In other words, today's psychologists apply diagnostic principles that are intertwined with the very same capitalist realist assumptions that permeate society at large.

Even that small segment of modern psychology that explicitly self-identifies as anti-capitalist has had difficulty avoiding the capitalist realist stereotype. 'Critical psychology', which challenges the limitations of mainstream psychology by emphasising the role of societal power structures in shaping human behaviour, ironically owes its academic existence to market forces in education. As outlined by critical psychologist Ian Parker, the field really began to establish itself in UK universities only after the deregulation of British higher education in the 1990s, when polytechnic colleges were given university status and encouraged to develop novel courses (Parker, 2009). The unleashing of academic entrepreneurialism empowered a latent eagerness to challenge traditionalism and embrace change, so long as such subversion could be commodified into saleable degree programmes. Private-sector academic publishers found themselves suddenly able to produce new journals and textbooks that leveraged this fresh niche market for critical psychology, helping to make the field more viable (or not, as the case sometimes was, with a number of critical psychology programmes eventually winding down due to a lack of student demand).

But despite the fact that critical psychology ostensibly seeks to challenge the inevitability of capitalism, it nonetheless perpetuates psychology's overall approach of depicting human beings as beacons of individuality, whose personal experiences and needs are held to be sovereign. Critical psychology's research methods frequently emphasise subjective perspectives as encapsulated in interview transcripts and other qualitative testimonies, which are said to enjoy a pre-eminence above all other sources of information. This 'textual empiricism' encourages the inference that nothing outside a text needs to be examined. Although critical psychologists draw attention to societal and political issues, their obsession with such pure forms of participant subjectivity propels their scholarship towards individuocentrism, and generates accounts of the human experience that are located within social contexts but nonetheless disconnected from them.

Critical psychologists themselves correctly point out that academia is ensconced within the capitalist system and as such is never likely to deviate too far from capitalist values. It is simply not in the interest of a capitalist society to subsidise ideas that negate the practices of capitalism. Even the so-called 'woke' university sector has succumbed to Fisher's concept of the 'business ontology', in which everything in society, including education, must now be run as a business (Fisher, 2009). Universities operate their own internal forms of market economy, where academics on individualised contracts are remunerated on the basis of how much their work-products benefit their corporate employers in terms of financial success and business rankings. In the so-called 'publish-or-perish' arena, the requirements of contemporary capitalism drive the norms of academic practice. The dominant forms of all academic fields, including and in particular psychology, will inevitably reflect these pressures. Psychology views human interaction and well-being in terms of capitalist norms in no small part because psychologists themselves are required to do so in their daily lives. It should be no surprise that academic psychology sees the dynamic of Western, neoliberal, and capitalist societies as reflecting the default mental processes of humankind.

17

The Capitalist Denial of Illness

In October 2020, during the first phase of the COVID-19 pandemic, the British prime minister was presented with a briefing paper describing the dangers of long COVID, some three months prior to George Monbiot's article in *The Guardian*. 'Early indications are that long-term effects include lung fibrosis, myalgia, tachycardia, fatigue, "brain fog", anxiety and depression,' the document's authors advised. 'Bollocks!' the prime minister scrawled with his biro onto the margin of the page, before adding: 'This is Gulf War syndrome stuff,' a reference to a condition common in war veterans that some doctors felt was caused by mass hysteria (Osborne, 2023). It was as if the prime minister—a conservative politician with no medical training—possessed an unstoppable instinct regarding what this novel malady really involved. So much so that even in the midst of what was an epochal global health emergency, he felt confident enough to reject the recommendations of his own scientific advisors.

Medical research would later reveal that Gulf War syndrome is *not* caused by mass hysteria, but by exposure to sarin nerve gas (Cruz-Hernandez et al., 2022; Haley et al., 2022). In fact, the claim that Gulf War syndrome was psychogenic had never been supported by affirmative research evidence, as research had for decades failed to identify *any* specific cause for the condition. But it was this very absence of knowledge that led some observers to theorise about psychological factors. The psychogenic approach was popularised by several media commentators, especially in the right-wing press. Many writers felt that scientific uncertainty about the syndrome's precise cause constituted 'a powerful motivation for leaping on a diagnosis', seducing sufferers into spuriously laying claim to the affliction as a means of seeking emotional comfort (Aaronovitch, 2014). The issue was further clouded by suggestions from some psychiatrists that the syndrome arose from emotional distress, and that '[media] misinformation about Gulf War syndrome' was therefore an important factor contributing to its prevalence (Mackenzie, 2004). This was precisely the narrative that would later be used to accuse George Monbiot of super-spreading long COVID.

Another echo with long COVID related to costs. Given the number of sufferers involved, medical treatment for Gulf War syndrome could eventually prove very expensive. And as sufferers were government employees who became sick in the course of their duties, there was also the issue of employer liability. All this impacted on how the powers-that-be responded to the condition. As one Member of Parliament testified to the independent inquiry that investigated Gulf War syndrome, the British government would ordinarily be 'very reluctant to admit responsibility for something which is going to involve them in writing large cheques' (Mackenzie, 2004). To this day, the UK government's website states that veterans who fell ill 'as a result of their service in the Gulf' are eligible to apply for a 'war pension'; yet, on the very same webpage, it is stated—as fact—that 'the current available research still does not yet identify a specific cause' for Gulf War syndrome (Office for Veterans' Affairs, 2023). Given that causality is officially deemed to be a *de facto* mystery, quite how veterans are expected to demonstrate that they became ill 'as a result of their service' is left somewhat unclear.

According to the website, the government's position arose from a meeting of Gulf War syndrome 'experts' convened by the relevant minister. It is very notable that of the eight experts listed, four of them were a professor of mental health, a researcher on 'combat stress', a director of mental health research, and a professor in 'the history of medicine and psychiatry' (Office for Veterans' Affairs, 2023). In other words, half of the government's 'experts' were psychologists, psychiatrists, or social scientists (if we generously include the historian in that final category). This perfectly exemplifies a recurring problem with how psychologists lay claim to medical knowledge in public policy forums. The dominant approach in academic psychology is to fetishise psychogenic theories of disease. This not only appeals to advocates of fiscal austerity, it also reinforces right-wing reluctance toward subsidising the unwell. Psychological ideas are welcomed by conservative opinion not least because they provide a means for debunking the benefit claims of the disabled. For their part, psychologists often appear delighted with positive political attention, and some of them seem especially keen to insert themselves into public debate—even (or especially) if it means displacing actual biomedical expertise.

The psychologising of physical disease is a seductive habit to which mainstream psychology has long been addicted. Its compatibility with cutbacks culture makes it attractive to conservative politicians, so much so that even world leaders (at least one of them) are occasionally wont to advocate on behalf of psychogenic hypotheses. But more importantly, in social justice terms, the psychogenic approach to illness is inherently regressive, despite the supposed 'liberal' bias in psychology. Its reliance on ableist, classist, and often sexist stereotypes places it firmly at the centre of the conservative worldview.

Illness Denialism

Long COVID, ME/CFS, and Gulf War syndrome are far from the only conditions to have been portrayed in this manner. In fact, during the initial months of the coronavirus pandemic, even incidents of acute-form COVID-19 itself were attributed to psychogenic causes, and not just by anti-lockdown conspiracy theorists on internet message boards. In May 2020, around six weeks after stay-at-home orders were first issued in the UK, the country's Joint Biosecurity Centre published a note advising government that some COVID-19 outbreaks might in fact be the result of 'local episodes of mass psychogenic illness' (Barnes, 2020), effectively suggesting that COVID-19 was not as rampant as it appeared from the published disease surveillance statistics. No evidence was supplied to support the claim that significant numbers of people who reported having COVID-19 were simply imagining it, and over subsequent months, as hospitalisations and fatalities spiked, it became apparent just how virulent and deadly COVID-19 actually was. Nonetheless, by apparently downplaying the severity of the pandemic in its early stages, the Joint Biosecurity Centre's logic was exactly the narrative that would later help to legitimise stereotypes regarding the validity of *long* COVID.

Claiming that at least some cases of COVID-19 were attributable to psychological hysteria is akin to claiming that the actual disease is less prevalent than epidemiologists would have you believe. Such an argument has many political implications. It undercuts the case for locking down the economy or for applying other protective measures that interrupt capitalist activity. It relieves conservative politicians of blame for having previously cut back on public health spending. It reduces pressure for new public spending on medical supplies and hospital services. And more subtly, it is convenient in culture-war terms: it feeds into a broader narrative that ridicules the political left for its stance on all these matters. It suggests that citizens are simply being naive if they submit to fear of physical threat or if they think that sacrifices might be required so that others can be helped.

Ultimately, downplaying the prevalence of COVID-19 was a form of pandemic denialism. Such forms of reality scepticism—in common with climate change denial, evolution denial, or even Holocaust denial—involve levels of conspiratorial thinking that reflect more than just a poor grasp of empirical principles. They would not exist without partisan political motivation. In the dynamics of political tribalism, the default posture for conservatism is to deny the existence of tangible dangers that demand collaborative responses or which threaten to otherwise undermine the status quo. Hence conservatism's preference to believe that conditions such as COVID-19—or, more recently, long COVID—constitute exaggerated, rather than real, crises.

18

'Personality Is Bad for You'

Psychologists have long sold their field as one that offers insights into the origins of physical disease. Up to and including terminal cancer, they have argued, ill-health is not just experienced psychologically—it is intertwined with psychological determinants.

This logic was notoriously exploited by the tobacco industry, which, for thirty years, sponsored some of academia's most esteemed psychologists to flood the literature with research purporting to show how personality and stress, rather than smoking, was what made tobacco users sick. The renowned British psychologist Hans Eysenck (who, as we saw in Part II, authored the eugenicist screed *The Inequality of Man*) received almost one million pounds of secret Big Tobacco funding in just over a decade (Pringle, 1996). Eysenck duly reported data suggesting that 'negative' personality traits were strongly predictive of both onset and progression of cancer and heart disease (Grossarth-Maticek, Eysenck, & Vetter, 1988a) and that psychological factors were as destructive to health, if not more so, than smoking itself (Eysenck, 1991). He also reported that psychotherapies designed to neutralise such traits could be used to prolong the lives of

© The Author(s), under exclusive license to Springer Nature Switzerland AG 2025
B. M. Hughes, *Psychology's Quiet Conservatism*,
https://doi.org/10.1007/978-3-032-07724-0_18

terminally ill patients by, for example, staving off cancer (Eysenck & Grossarth-Maticek, 1991).

Eysenck even published data suggesting that *being opposed to smoking* was more dangerous than *being a smoker*—literally: that anti-smokers die younger. According to Eysenck, his data revealed that people who express hostility toward tobacco exhibited a toxic form of prejudice equivalent to that seen in racists and other bigots (Grossarth-Maticek, Eysenck, & Vetter, 1988b). It was this toxicity that was said to lead to premature death.

Although the source of Eysenck's research funding was kept secret for many years, recurring questions were raised about the quality of his methods and the integrity with which he analysed and reported his data. After complaints were formally lodged in 2019 (Marks, 2019; Pelosi, 2019), dozens of Eysenck's papers were investigated, found to be unsound, and officially retracted. Eysenck's claims that heart disease and cancer were shaped more by psychogenic processes than by tobacco (along with his assertion that anti-smokers were terrible people) were entirely discredited.

None of this is to say that psychological factors are irrelevant to physical disease. Psychological factors influence a person's behavioural choices, such as whether they choose to smoke in the first place. In that sense psychological factors can certainly be said to contribute to physical health outcomes. In addition, emotional challenges prevent the human body from achieving states of restfulness. Constant anxiety impedes appetite and disrupts sleep, both of which in turn exert their own impacts on human health. It is also well established that chronic stress elevates blood pressure, which in and of itself is simply unhealthy. There is plenty of evidence to show that the heightened blood pressure associated with sustained stress exposure leads to physical wear-and-tear on the cardiovascular system, akin to when an engine risks malfunction after being continuously over-revved (see, for example, Whittaker et al., 2021). Chronic stress is also associated with the disruption of various immune system processes, again most likely due to the way stress provokes physiological responses in the body.

However, the fact that psychological stress is statistically associated with high blood pressure, and so in turn is related to a heightened risk of eventual atherosclerosis, does not negate all the *other* factors that are known to engender cardiovascular disease, most of which have far more direct effects. We know that stress is clinically relevant, but we also know that cardiovascular health is much more likely to be damaged by smoking, sedentary lifestyles, high-fat diets, excessive bodyweight, drug misuse, caffeine consumption, intake of sodium (i.e., salt), environmental pollution, medication side-effects, bacterial infections, viruses, comorbid conditions (chiefly diabetes), and many other similar dangers. Cardiovascular disease processes are also subject to genetic factors (family history is a strong predictor), gender (men are more likely to get heart disease than women), and the physiological consequences of ageing (older people's hearts are just weaker). In short, lots of things lead to heart disease. Even if some of the causal factors are behaviourally mediated, most are simply non-psychological. When a person develops heart disease, the likelihood is that one or more of these non-psychological issues will have proven decisive. Overall, the role of psychological factors in physical disease is almost always peripheral, marginal, or incidental. Any claim to the contrary is simply an exaggeration.

While psychologists can offer useful input into the science of stress biology, their theories and methods are inconsequential in isolation. The belief that psychological factors somehow *dominate* causal mechanisms, that they *determine* whether someone becomes sick, results from reductionist attitudes and an enslavement to individuocentrism. Theories that the human body is capable of falling ill solely because of psychological forces verge on the fantastical. Moreover, they distract attention from the societal and economic root-causes of population ill-health. As we discussed in Part II, the most important contextual issues are usually far beyond what psychology can deal with. Economic impoverishment is probably the most important predictor of human ill-health that we know of. When it comes to reducing population heart disease, for example, neither biomedicine nor psychology will ever be as effective as a political intervention that successfully reduces poverty.

We should also note that disease-inducing levels of poverty are experienced mostly by social and cultural out-groups. Disproportionate numbers of minority communities and marginalised subcultures face the direst economic challenges. Women too are discriminated against in ways that create for them economic deficits and barriers. This then imbues psychology's reasoning with a sinister tone. If we follow the emphasis of mainstream psychology, we should not discuss the ill-health of women or minorities in terms of the poverty, exclusion, or discrimination that they face; we should instead approach it in terms of their cognitions, emotions, choices, and psychological traits. The message from psychology, albeit implicit, is stark. When it comes to being sick, people largely have *themselves* to blame.

19

The Psychologising of the Sick

Psychologists have spent a century devising theories of physical ill-health. The emergence of ideas tends to follow a recurring pattern. Just like with long COVID or Gulf War syndrome, each malady starts by being 'medically unexplained'—either the research has not been done or the scientists cannot agree on what the data mean. This then presents an explanatory gap into which psychogenic theories can be inserted. Even after the physiologists, virologists, microbiologists, haematologists, cardiologists, immunologists, and toxicologists have given up on a problem, psychologists (and their psychiatrist cousins) often appear happy to promote their own one-size-fits-all explanations. For them, ambiguous data and medical uncertainty are not seen as clouding the issue. Instead they are considered to constitute evidence *in favour* of psychogenesis: whenever an illness cannot be explained, it is *presumed* to be psychological.

Perhaps the archetypal example of this reasoning is the belief that stomach ulcers are triggered by psychological stress. For many decades advocates of psychogenic theories argued that peptic ulcers were produced by emotional turmoil, which was said to create excessive acid secretion in the stomach. Even the pioneering gastroenterologist Burrill Crohn, the co-discoverer of Crohn's disease, considered it an 'accepted fact' that ulcers were rooted in 'psychic strain' (Crohn, 1947). Doctors were persuaded that ulcers arose after patients faced unconscious psychic conflicts (Fordtran, 1973) or periods of 'emotional turbulence' (Castelnuovo-Tedesco, 1962). It was recommended that practitioners use this 'knowledge' when diagnosing their patients. Medical students were taught that typical ulcer patients could be identified by their 'anxious, lined face' and constant 'alertness and awareness' (Illingworth, 1956).

But by the end of the twentieth century, it was discovered that nearly all ulcers were in fact caused by the bacterium *Helicobacter pylori*, which weakens the stomach's mucosal defences against acid (for a small but significant subset of patients, these defences can also be degraded by the heavy use of certain medications; Yeomans, 2011). Without these damaging physical processes, no amount of psychological stress could ever produce a stomach ulcer in the human body. Once the precise role of *H. pylori* was established, there was no longer an explanatory gap for psychologists to exploit. Antibiotics replaced psychotherapy as the primary recommended treatment. The psychogenic ulcer theory was left behind.

However, the psychosomatic paradigm has not always shifted so neatly. Psychogenic theories continue to abound. Virtually every system of the body has been included in the purview of psychogenic speculation. The list of psychologised medical conditions is therefore extensive.

'The Faker's Disease'

Despite being a major cause of physical disability, multiple sclerosis, or MS, has a long history of being cast as a psychogenic condition. We understand today that MS is a complex autoimmune disease, in which the immune system starts to attack the myelin sheaths that ordinarily protect the body's own nerve cells. But such precise medical knowledge has emerged rather slowly. Effects on the brain were not visible until the invention of magnetic resonance imaging in the 1980s (Gurfinkel, 2024). And it was not until the 2020s that researchers demonstrated how MS can be triggered by viral infection (Lanz et al., 2022). This history of clinical uncertainty allowed an explanatory vacuum to persist, into which numerous psychological tropes have proliferated.

At first MS was viewed as a form of malingering, with the condition being known to many physicians as the 'faker's disease' (Weintraub, 2009). The concept of an 'MS-prone' personality type emerged more than a century ago. The disease was said to be rooted in patients' emotional immaturity (LaRocca, 1984). In the 1920s, physicians attributed MS symptoms to a person's failure to deal with emotional distress, and even to their 'oedipal fixations', a then-fashionable psychoanalytic idea where young men are suspected of being sexually attracted to their mothers (Jelliffe, 1921). After doctors came to appreciate that MS engendered stark physical changes in the human body, many persisted in framing its cascading physical decline as being triggered by emotional damage. A litany of research papers purported to report evidence that MS patients had histories of 'personality problems', 'poor emotional adjustment', and 'neurotic difficulties' (Langworthy, 1948). Even in the twenty-first century, papers in psychology have linked MS to 'early childhood trauma' and 'a lack of positive bonding experiences' in infancy (Munzel, 2002).

As MS is a degenerative disease, its initial symptoms—which include fatigue, dizziness, blurred vision, and poor concentration—can often seem ill-defined, if not even inconsequential. Unfortunately, this vagueness only further opens the door for psychogenic theorising. When the renowned British concert cellist Jacqueline du Pré first showed symptoms of MS in the early 1970s, she was told not to worry, that they were merely signs of depression, stress, and adolescent trauma (Mezaki & Namekawa, 2019). One doctor informed her that the numbness she felt on the soles of her feet was caused by 'unconscious resistance' (du Pré & du Pré, 1997). Jacqueline was made to wait years for a proper clinical diagnosis, which only arrived once her physical disabilities became so pronounced as to be unignorable. Within months of being diagnosed with MS, her prodigious musical career was cut short by illness. She then lived through fourteen years of further physical deterioration before dying in 1987 at the age of 42.

MS patients continue to encounter scepticism. When Hollywood movie actor Selma Blair first developed signs of MS, her doctors advised her that she was burned out from overwork. They told her directly that her fatigue and dizziness were unsurprising given that she was a 'single mother' (Thorbecke et al., 2019). At one medical appointment, where she had complained of chronic pain, a doctor recommended that perhaps she 'needed a boyfriend' (Clark, 2023). Blair had starred in several hit movies, including *Cruel Intentions* and *Legally Blonde*, but even with her celebrity status and wealth, she was unable to access healthcare providers who could produce a timely diagnosis. According to her own estimation, she 'probably had this incurable disease for fifteen years at least' before doctors started to take her symptoms seriously (Lee, 2019).

Perfectionism Pain

Another frequently psychologised disease is migraine. Unlike MS, migraine is not life-threatening or degenerative. However, it is nonetheless an especially unpleasant neurological disorder. Sufferers experience severe episodic headaches, sometimes accompanied by vomiting, dizziness, and other adverse symptoms. Around a third of sufferers report a type of perceptual disturbance known as an 'aura', a phenomenon caused by depolarisation in the brain cortex that is also seen in epilepsy (where it is understood to constitute an actual seizure, albeit a minor one). The success of certain types of surgery suggest that migraine headaches might well be caused by the specific features of cranial anatomy that a person happens to be born with, helping to explain why migraine often runs in families.

However, specific insights about the physiology of migraine have emerged only recently. For many years, even centuries, migraine was treated as a medical mystery. Once again, psychologists filled the explanatory vacuum with a plethora of causal theories. A crucial point is that migraine has always been about twice as likely to arise in women as compared to men. As such, gender has always loomed large in migraine lore.

In the 1930s, psychologists explained that migraine arose from the inherent frailties of femininity. These included women's 'exaggerated sense of personal insecurity', as well as their 'perfectionist traits' and 'tendency to anxiety and worry unduly' (Touraine & Draper, 1934). A particular threat was said to be women's unreasonable impatience to succeed in the (male) world of work. 'Ambitious' young women were believed to be particularly at risk (Knopf, 1935). Women who fell short of stereotypical womanliness—at least as deemed in terms of male preferences—were also seen as susceptible. A common claim was that migraine sufferers were much less likely to experience 'a successful heterosexual life' (Touraine & Draper, 1934).

Over the twentieth century, theories of migraine evolved. Migraine came to be viewed not just as a byproduct of negative personality traits, but as a sign of full-blown mental illness. Psychologists argued—without evidence—that women who suffered migraines had more 'psychiatric disability' than men (Pearce, 1977). These women were depicted as psychologically primed for self-destruction, as bringing the condition upon themselves. Migraine was seen as 'the price [women] pay for self-induced fatigue.' In severe cases, it was considered to be part of 'an attitude of dependency, [characterised by] demands for attention or excuses for escape from responsibilities.' Not only were women cast as seeking sanctuary in self-induced ill-health, they were also accused of cynically exploiting the gullibility of those around them. According to psychological theories, it was 'not uncommon' for women who had migraine to engage in 'hysterical manipulation' of their families, friends and workmates (Pearce, 1977).

Even after researchers established the neurological mechanisms of migraine, psychologising stereotypes have remained in circulation. Well into the twenty-first century, psychologists continue to report survey data linking migraines to women's 'locus of control' (Furnham & Cheng, 2023), 'income dissatisfaction' (Rondinella & Silipo, 2023), and 'catastrophising' tendencies (Farris et al., 2020). The only purpose in reporting such findings is to implicate these attributes as probable causes, or accelerators, of migraine. But such interpretations involve drawing arbitrary inferences from correlational data. They are framed as if a woman's sense of helplessness, dissatisfaction, and pessimism must somehow *cause* pain—when common sense suggests they are just as likely, if not more likely, to *be caused by* pain. Similarly, personality traits continue to lurk in the migraine literature. Migraine patients are frequently reported as exhibiting high scores for traits such as 'rigidity' and 'obsessivity' (Sánchez-Román et al., 2007) and 'social introversion' (Karakurum et al., 2004). Once again, these traits are presented as contributing to the onset of migraine. It is as though the researchers cannot see how chronic episodic pain might affect the way patients respond to personality questionnaires. It should be unsurprising that a life of daily pain leads people to describe themselves as constrained, preoccupied, or reclusive. These self-reported traits are not the reason for the presence of migraine; the presence of migraine is the reason for these self-reported traits.

Migraine is difficult to cure, but it is generally understood that the best treatments are pharmacological (Eigenbrodt et al., 2021). Prevention too is best achieved using drugs such as beta blockers and other blood-pressure-lowering pills. But for more than a century the psychogenic view served to normalise the demeaning of migraine sufferers by portraying them as emotionally flawed, while affording scientific credibility to insidious sexist stereotypes. There is little doubt that psychological theories of migraine have thus led to much unnecessary suffering. And yet the psychologising of migraine goes on. It is as though psychological theories maintain their currency not *in spite* of their pejorative impact, but *because* of it. Psychogenic narratives ring especially true to audiences who agree that many of the people complaining of migraine symptoms are not *really* that sick—instead they are mentally weak, self-destructive, and, of course, female.

Capitalist Psychosomatics

Examples such as stomach ulcers, MS, or migraine show how the psychologising of disease shifts blame away from society and culture, and onto sufferers themselves. A recurring theme is that psychogenic theories are presented without clear confirmatory evidence but are none-theless used to fill an explanatory vacuum. A more subtle feature is that all such theories serve political purposes. They share a common valence in the culture wars.

Blaming ulcers on sufferers' own fragility draws attention away from the excessive demands of the capitalist economy. According to this view, people who become sick because of workplace stress are simply too weak to survive (and so should not be there in the first place), or else are just malingering (and so should be ignored, or worse, punished). Similarly, suggesting that women experience migraines or MS because they are emotionally disabled serves to reinforce conservative stereotypes about female timidity. And claims that women become ill due to excessive professional 'ambition' reflect a longstanding bias against female eco-nomic empowerment. The lesson being projected is that women can (and should) avoid the risk of disease by striving to be *less* ambitious; in other words, by withdrawing from the boardroom and sticking to their more 'natural' domesticated roles. Nowhere in these narratives is it acknowledged that people might become ill because of societal negli-gence, degradation of public services, or toxic economic demands.

By contrast, biological explanations that ulcers result from bacteria, that MS is triggered by infections, or that migraine is caused by anato-mical abnormalities shift responsibility back onto wider society. Instead of blaming individuals for their own illnesses, the issue becomes one of how those who fall ill in our societies can be supported to live lives of maximum dignity and minimal suffering, and of how those who are *not* ill can be protected from disease by ensuring they have clean environ-ments, adequate nutrition, secure social safety nets, access to healthcare, and freedom from injustice, violence, and other toxic experiences throughout their lives. Biological insights help to highlight the ecological interconnectedness of humanity and our collective responsibilities to one another. They challenge the dogmas of capitalist realism that talk of people being free to live their own lives, but only so long as *other* people can be blamed for dying *their own* deaths.

Absence of Evidence

Ulcers, MS, and migraine are just three medical conditions that have been depicted by psychologists as being rooted in psychological causes. In reality, as mentioned above, literally hundreds of diseases have been psychologised in this way. Gulf War syndrome and long COVID are two recent examples. Other diseases that have at one time or another been seen as partly or entirely psychogenic include arthritis (Johnson et al., 1947), tuberculosis (cf. Dormandy, 2000), asthma (Graham et al., 1967), diabetes (cf., Dunn & Turtle, 1981), epilepsy (Clark, 1926), irritable bowel syndrome (Sykes et al., 2003), chronic Lyme disease (Hassett et al., 2009), celiac disease (Prugh, 1951), hypothyroidism (Lidz & Whitehorn, 1950), psoriasis (Wittkower, 1946), endometriosis (cf., Freedman, 2017), and even AIDS (Schmidt, 1984).

But alongside disease misclassification, there is the problem of symptom misdiagnosis. The medical and scientific literature is replete with cases of individual patients whose symptom reports were initially dismissed as signs of psychological frailty, only for it later to be discovered that in fact they were caused by serious medical illness (Goudsmit, 1994). Examples abound: a woman whose stomach cramps were attributed to workplace stress, when they were actually caused by cancer (Brozovic, 1989); a woman who was referred for a psychiatric evaluation after turning up at a hospital emergency room with stroke symptoms (Yong, 2020); multiple women whose diabetic dyspnoea was dismissed as 'hysterical hyperventilation' (Treasure et al., 1987); and the four-out-of-five disabled British women and non-binary people who, in a national survey, reported how doctors initially attributed their debilitating physical symptoms to anxiety, stress, and other psychosocial causes before only later arriving at proper medical diagnoses, delays that in a third of such cases lasted more than a decade (Graham, 2024).

In 2022, courts in Ireland dealt with a case involving a young man whose brain tumour symptoms had been dismissed by hospital doctors as 'psychological'. At the age of 14, the boy started to develop severe headaches, fatigue, and even partial paralysis, but his local hospital insisted on referring him to mental health services. The tumour slowly growing in his brain was revealed only after his mother arranged for an MRI scan to be conducted at an independent private healthcare facility. The boy underwent immediate surgery but, as the initial psychogenic diagnosis had delayed everything by five months, he was left with life-long deficits. While maintaining that their actions had not impacted the boy's medical outcomes, the hospital nonetheless had little choice but to admit negligence. The courts approved a compensation settlement of over €6 million for the child and his family (O'Loughlin, 2022).

Inflated faith in the psychogenic hypothesis is not just a historic problem, but part and parcel of contemporary psychology. Its problematic presumption—that *whenever an illness cannot be explained, it must be psychological*—is embedded within professional clinical practice in mental healthcare. The current International Classification of Diseases (ICD-10), the World Health Organisation's official directory of medical classifications, presents 'unexplained somatic symptoms' in its section on 'psychiatric disorders'; effectively guiding doctors to interpret the *absence* of evidence of specific biopathology as equating to the *presence* of evidence of a psychiatric issue (O'Leary & Geraghty, 2021).

In the DSM-5, the American Psychiatric Association's influential counterpart to the ICD-10, doctors are advised to consider 'medically unexplained' symptoms as grounds for a diagnosis of so-called 'conversion disorder' (or 'functional neurological symptom disorder')—again, a psychiatric malady. Crucially, a doctor in this situation need not establish whether any specific 'psychological factors are judged to be associated' with the unexplained symptoms (Scamvougeras & Howard, 2020). The mere absence of a demonstrable physical pathology is said to be grounds alone for diagnosing the patient as psychologically disturbed. Given such guidance, it is perhaps unsurprising that even in the 2020s, many doctors have greeted the emergence of long COVID—for which no particular physical causal mechanisms can at present be confirmed through medical testing—as a possible, if not indeed probable, psychological event.

20

Unidentified Psychic Objects

The central problem with the psychogenic approach is encapsulated in astrophysicist Carl Sagan's famous logical aphorism: 'absence of evidence is not evidence of absence' (Sagan, 1997). Just because doctors cannot locate the physical cause of a patients' symptoms does not mean that no such physical problem exists. The absence of evidence of a specific physical pathology is not evidence of the absence of such pathology. Moreover, it certainly is not evidence that the true cause must therefore be psychological.

The history of medical science demonstrates how incomplete knowledge frequently presents doctors with mystery illnesses that at first defy understanding. Only after researchers have had time to identify the actual physical mechanisms involved can patients' symptoms duly be 'explained.' Until this is done, any such symptoms must necessarily, by definition, be considered 'unexplained'. Hasty claims that they are psychological are at best premature, and at worst nonsensical.

In short, when psychologists assert that unexplained physical symptoms *must have* psychological causes, they are committing a rudimentary error of logic. Their claims resemble those of conspiracy theorists who suggest that every 'unidentified' flying object *must be* an alien spaceship.

© The Author(s), under exclusive license to Springer Nature Switzerland AG 2025   **151**
B. M. Hughes, *Psychology's Quiet Conservatism*,
https://doi.org/10.1007/978-3-032-07724-0_20

In reality, most UFOs, while mysterious at first, are eventually found to be prosaically explicable, often turning out to be high-altitude surveillance balloons, unusual atmospheric phenomena, or even birds. For the small number of UFOs that are never explained—where the video footage is too fuzzy or where eye-witness accounts are uncorroborated—it simply does not make sense to interpret this lack of certitude as *itself* constituting evidence of extraterrestrial intelligence. In precisely the same way, the history of medicine shows us that hundreds of so-called 'mystery' illnesses have arisen over the years, most of which are eventually explained after the necessary research has been completed (even if that process occasionally takes decades). There is no basis ever to conclude that 'unexplained' symptoms must result from mental processes, simply by virtue of *being* unexplained.

The practice of interpreting unexplained physical symptoms as evidence of psychiatric dysfunction is exceptionally weak. For one thing, such a diagnosis can never be definitive: symptoms that are unexplained now might turn out to be explained later. Logically speaking, it is never possible to exclude such an eventuality. A second issue concerns the mechanics of diagnosis. The logic inherent in arriving at a diagnosis of 'medically unexplained symptoms' is that it results from a process of elimination: the clinician must first exclude *all* other possible diagnoses. Given the hundreds of thousands of different illnesses that any single patient might have, such a process of elimination—if pursued methodically—would take an extremely long time. In reality, however, this type of sifting is never conducted. It is, after all, a practical impossibility.

Instead, physicians invoke their intuitions and take account of signs and signals to help them through their deliberations. In one study where researchers asked doctors to produce diagnoses based on video-recorded medical consultations, the doctors were shown to produce their diagnoses of 'medically unexplained symptoms' in a matter of minutes (Houwen et al., 2020). For patients who had long-term symptoms, the doctors took an average of less than *fifteen seconds* to do so. In other words, it took them less than fifteen seconds to exclude hundreds of thousands of other possible medical conditions. In many cases, the doctors formed their diagnoses of 'medically unexplained symptoms' before the patient had even started talking. One thing was clear: these doctors were not engaging in a true process of elimination. Rather, they seemed to be diagnosing patients (at least partly) on the basis of their appearance, demeanour, and other nonverbal cues. In other words, they came to their conclusions by jumping to them.

The Individual (Illness) is Sovereign

The field of 'medically unexplained symptoms' research is dominated by psychologists and psychological theories, and is premised on the assumption that an unexplained illness must be a sign of mental dysfunction. Its apparent weddedness to this fallacious psychosomatic reasoning makes the field of psychology far from 'liberal'. In fact it makes it the opposite: the theories that thrive in this domain of academia are fundamentally grounded in socially conservative worldviews. By insisting that psychological factors are pivotal in precipitating physical pathology, the field of psychology presents a classically conservative manifesto—explanations that emphasise the primacy of personal responsibility and which propose a largely asocial basis for human suffering. Psychology's habits of reductionism and individuocentrism operate to imply that psychometric test scores and superficial demographics (such as age or gender) are all we need to explain physical demise.

The natural orientation of psychology is to argue that psychosomatic mechanisms are important contributors to physical disease. That the mind is as powerful as it is mysterious. It is as if human beings, should they be placed under sufficient emotional duress, are liable to keel over with (literally) broken hearts. This is complemented by a near-Nietzschean narrative of psychological undercurrents driving humans toward their own destruction. Once trapped within a vicious cycle of self-defeating cognitions and behaviours, the narrative goes, only psychological intervention can rescue the afflicted client.

The victims of illness are to blame for their own victimhood. The impact of other people—the ruling classes, privileged in-groups, and those who turn blind eyes to societal problems such as poverty and marginalisation—is ignored. The shifting of focus away from the power-ful and onto the personality traits, poor choices, or 'illness beliefs' of those who fall victim to life's worst misfortunes, is exactly in line with the doctrines of capitalism to which neoliberal conservatives would like us to believe there is no alternative. All the better if victims can be stigmatised as benefit scroungers or malingerers, as this keeps the masses fighting among themselves, thus staving off revolution.

The promotion of psychology as the answer to all that ails you panders to the priorities of conservatism. This is true regardless of how many individual psychologists identify as socially liberal. Most psychol-ogists remain oblivious to (or uninterested in) the political biases embedded in their psychosomatic, psychogenic, and biopsychosocial theories. Practitioners' own politics influence mood rather than method. Personal views matter little when all the journals, textbooks, and training manuals are stacked with conservative talking points.

21

Pathology and Protectionism

It was something of a coincidence that the COVID-19 pandemic over-lapped with the final stages of a regulatory controversy in Britain con-cerning that country's standards for the treatment of ME/CFS. In October 2021, the National Institute for Health and Care Excellence (which styles itself as NICE) finally published its new guideline for dealing with the disease, replacing the previous guideline which had stood since 2007. Medical science had developed considerably since the previous version was published, and the new standard departed substan-tially from what had previously been proposed (Marsh, 2021). Following a detailed review of hundreds of research studies, NICE was now advising doctors that ME/CFS was a complex neurological condi-tion and that psychotherapy should therefore not be considered as a valid frontline treatment. In addition, NICE stated that doctors should no longer use behavioural exercise interventions intended to train patients to 'push through' their conditions (such 'graded exercise thera-pies', NICE pointed out, were in fact *damaging* to patients' welfare). Effectively, NICE was confirming that ME/CFS should no longer be considered a psychogenic condition. Instead, its complex biomedical features would require personal care plans individualised for each

B. M. Hughes, *Psychology's Quiet Conservatism*,
https://doi.org/10.1007/978-3-032-07724-0_21

sufferer, based on inputs from neurologists, rheumatologists, endocrinologists, immunologists, and infectious disease specialists. Psychologists would only contribute, from the sidelines, if a particular patient was to present with an emotional support need.

The accumulating research literature had turned the tables on the psychogenic model of ME/CFS. The original 2007 guideline had been based on just nine studies examining psychotherapy-based interventions for ME/CFS. By contrast, the 2021 guideline benefited from a vast library of some 230 findings from newly conducted research trials (Hughes, 2021). Within these hundreds of new studies, NICE's reviewers could find no evidence to support the use of cognitive behaviour therapy or graded exercise with ME/CFS patients. In many respects, these approaches were found to be harmful. Not only were thousands of patients receiving no tangible benefit beyond possible placebo effects, many physically deteriorated after receiving the treatments. Psychological therapies were not merely useless, but literally *worse* than useless.

Psychology is, of course, an evidence-based science. So you might expect that psychology practitioners would have been grateful for this substantial, evidence-based update to the ME/CFS guideline. However, this was not the universal reaction. The same voices that lambasted George Monbiot for describing the physical symptoms of long COVID in a national newspaper were now suddenly aghast that NICE was outlawing their preferred treatment approaches for ME/CFS, a condition that had provided psychologists with tens of thousands of clients in the UK alone over several decades. The various bodies representing psychological practitioners (including psychiatrists) came under pressure to object, and organisations such as the Royal College of Psychiatrists and the Royal College of Physicians eventually submitted formal written complaints to NICE, querying why changes to the psychogenic approach were needed. NICE agreed to a formal consultation with the Royal Colleges, where the basis for the new guidance could be explained in person to them.

Intriguingly, the official record of that in-person meeting showed that the professional representatives mounted no argument about the science behind ME/CFS and its treatment. Instead they cited 'concerns that [the new guideline] may have a negative impact on commissioning of services' (NICE, 2021). In other words, the people providing psychological therapies were concerned that their services would no longer be needed. Neither the speciousness of their psychogenic theories nor the dismal research evidence for their therapies appeared to come into their thinking. These therapy providers simply wished to preserve the status quo, *regardless* of what science might have to say about what their therapies could achieve (or, more correctly, were *failing* to achieve).

Unsurprisingly, NICE was not swayed by this concern. And so the new guideline was formally released at the end of October, 2021. There then followed a brief flurry of old-guard commentary on social media, letters to newspapers, and at least one academic journal article bemoaning the new reality (Hall, 2023). But ultimately their campaign of resistance was to no avail. NICE responded by politely but methodically debunking each complaint (Barry et al., 2024). Their independent regulation of medical standards was based on a dispassionate analysis of peer-reviewed research. It would not be derailed by loud voices, special pleading, or status concerns.

Tragically, that same month, a 27-year-old woman in Exeter in the southeast of England died at home from natural causes—namely, malnutrition and dehydration—that a coroner concluded arose 'because of severe myalgic encephalomyelitis (ME)' (Smith, 2024). Maeve Boothby-O'Neill had been ill for thirteen years. Despite persistent symptoms, her doctors reacted on the basis that her illness was psychogenic. One told her directly that her symptoms were 'all in her mind' (Morris, 2024). Even after being formally diagnosed with ME, she continued to face resistance from care providers. At the inquest following her death, Maeve's father described how the family had persistently been obstructed by a 'medical orthodoxy' that believed ME to be 'a behavioural problem or a psychological illness' (Tuller, 2024). Maeve became housebound and was unable to look after herself. Even then, her family faced difficulty obtaining palliative care, because of a suspicion that her symptoms were 'invented' (Morris, 2024). Maeve died on October 3, 2021, just three weeks before NICE published its new treatment guideline.

'I feel her loss every hour of every day,' Maeve's father told the coroner.

Depoliticising Distress

Psychologists supposedly subscribe to what is known as the 'biopsycho-social' model of health and illness, a set of ideas that ostensibly integrates biological, psychological, and social dimensions. However, psychologists being psychologists, their emphasis is nearly always on psychological factors. The biopsychosocial model is treated not as a balanced inter-disciplinary integration of *bio, psycho,* and *social,* but as a way of shoe-horning the *psycho* into medical discourse. Although lip service is certainly paid to the other two elements, a glance at the contents pages of journals of health psychology quickly reveals a field in which the dominant considerations are of cognitive, behavioural, perceptual, and interpersonal (as opposed to societal) processes. The *bio* exists mainly insofar as (some) of the research participants are sick; the *social* is addressed with reference to such person-level notions as *social cognition, social support,* or *social identity.* The individuocentrism of the work is strongly apparent.

While ostensibly an academic initiative, the biopsychosocial model was formally appropriated into the UK's disability benefits system in the early 2010s (Shakespeare et al., 2017). Two civil servants adapted the model to craft a framework for deciding a citizen's entitlement to out-of-work disability payments. It is notable that the model's original huma-nitarian intentions were largely dispensed with in the rewrite. The government version of 'biopsychosocial' logic instead focused on ways to distinguish *real* incapacity benefit claimants from fake ones. In doing so, its authors made a number of contentious claims.

In their document setting out the reasoning behind the new system for assessing people's work capabilities, the civil servants asserted that 'biological, psychological and social factors … can influence the course and outcome of any illness' (Waddell & Aylward, 2010; p. 22). This framing is highly dubious, because it is tenuous to suggest that psychological factors can influence the outcome of *any* illness. They also suggested that 'capacity may be limited by a health condi-tion, but performance is limited by how the person thinks and feels about their health condition' (p. 20). Once again, this overstates the influence of psychological factors—a person's ability to perform a given job will depend on much more than just how they 'think and feel' about their health. Individuocentrism was also a problem. The civil servants ignored all macro-level socioeconomic factors that might intrude on a person's health. Instead, they appeared to operate from the assumption that people who seek disability support should generally be distrusted. Many are not even incapacitated, they noted, 'although this does not mean they are *all* malingerers or scroungers' (p. 5; emphasis added). Terms such as 'malinger' and 'scrounger' are deeply pejorative, judgemental, and prejudicial. They convey a presumption of guilt, a classist stereotype that portrays benefit claimants as probable liars. Overall, there is little to suggest that the biopsychosocial model was particularly 'liberal' in this instance (Shakespeare et al., 2017).

Right-wing commentary increasingly demeans people who have disabilities or chronic illnesses, such as when criticising expenditure on accessibility initiatives (Nović, 2025), mocking the use of sign language (Garcia, 2025), or defending the use of the word 'retard' (Kirkland, 2025). But left-wing commentary too is concerning when it emphasises the importance of 'working people', thereby reinforcing the trope that a person's value is determined primarily in terms of their ability to be economically productive (Pring, 2022). The overall stance is one of 'healthism'—the belief that maintaining an ability to work is nothing less than a person's civic responsibility (Cheek, 2008). Its consequent implication is that people who find themselves *disabled* are failing their fellow citizens, and so are deserving of less public solidarity or state support (Clifford, 2020). 'Responsibilising' people for their own well-being (Teghtsoonian, 2009) aims to return the sick to their supposedly rightful status—as productive citizens in neoliberal workplaces and marketplaces.

Psychology has largely internalised the convention whereby suffering and wellness are considered in market terms. The financial case for mental health services is usually framed in relation to potential benefits for the economy. Depression, we are told, should be treated (or prevented) because it is a leading cause of disability. It accounts for enormous amounts of 'lost productivity'. The argument is based on how depression damages the welfare of *other people*, primarily employers, rather than that of its own victims (Cosgrove & Karter, 2018). Its implicit deference to employer concerns is the epitome of capitalist realism.

The psychologised discourse on the global burden of mental illness serves a similar purpose. Initiatives such as the World Health Organisation's *Depression: Let's Talk* campaign—supposedly applicable to people all over the planet, regardless of their political circumstances—promote a distancing between people and the root causes of their predicaments. A 2017 headline on the WHO's mental health website read 'When sadness doesn't stop: Helping Syrians talk about depression', as though a priority for Syrian people living through what was then one of the gravest humanitarian and refugee crises of the twenty-first century would be to receive Western-styled mental health interventions such as talk therapy (Cosgrove & Karter, 2018). (One might be tempted to speculate about how many 'disability-adjusted life years' a typical war refugee might end up costing their employer.) The overall effect of psychology's discourse is to depoliticise human distress.

Psychogenesis and Politics

To cast ill-health as a mental product is to advance an argument for social conservatism. It is to justify keeping things the way they are. Diseases that emerge from (or which are located in) the 'mind' do not require society to change its ways, or, more prosaically, to redistribute its resources. If the poor end up ill through misperception or faulty choices, then the non-poor are blameless. Disempowerment, social marginalisation, or even war need not be seen as the source of public health crises. By extension, it can be argued that relative health disparities reflect naturally occurring—and thus *necessary*—societal hierarchies. There Is No Alternative. There is certainly no case for overhauling society. The best we can do is to promote piecemeal token changes in individual behaviour.

Apart perhaps from time, health is the most valuable asset any human being can possess. Indeed, individual health is a commodity: a life protected from the risk of disease does not come free, and healthcare itself is manifestly expensive. Health therefore correlates with socio-economic prosperity and societal privilege. It is a signifier of material wealth and social status. According to the UK's Office for National Statistics, not only can people who live in the richest counties in England expect to live a decade longer than those who live in the poorest, they can also expect to experience a further decade of superior physical health (Rea & Tabor, 2022). This is why human health is a political issue. The fact that health is not shared equally across society makes it a right that must be fought for, a front not just in the culture wars, but in the class war too.

Inequalities of wealth, and thus health, are so prevalent as to be impossible to ignore. Even the least-informed citizens of nations where food is plentiful will be obliquely aware that millions of their fellow human beings around the world face daily threats from poverty, war, and malnutrition. But the most pernicious comparisons occur locally rather than globally. Within each society, high-earners will be well aware of their low-paid neighbours. The cash-rich will be well aware of the poor. The employed will be well aware of the unemployed. And people with houses will be well aware of the homeless. Privilege, therefore, should come with a significant risk of guilt: how can we who are ethical but prospering justify our daily tolerance of the fact that others are so less well off than we are? The solution, of course, is to rationalise the overall state of things, a pressure to compartmentalise that is more than amply met by cognitive dissonance and other forms of psychological positioning.

Despite claims about the field's 'liberal bias', psychology's models of well-being are framed in ways that consistently produce—and protect—conservative conclusions. Its dominant emphasis on individual agency, rather than on system-dependency, emboldens conservative scepticism toward liberal solutions for what are societally shared problems. Psychology views capitalism as a feature of humanity, not a bug. Its assumptions are drawn from cultural narratives that see all inequities as ordinary—we consider it normal that we live in a world where we look up to *him* because he is upper-class, and down on *him* because he is lower-class. The diseased, we are told, are sick because of who they are; at best, passive victims of predestined misfortune; at worst, an untrustworthy underclass of malingerers and hypochondriacs, deludedly coddled by the woke.

Part IV

Modernity and Declinism

22

Generation Snowflake

These days, life comes at us very fast. The only constant, as the cliché has it, is change. Our worlds have become larger and more convoluted. Our daily lives would likely be unrecognisable to our grandparents, never mind to our earlier ancestors. While a typical medieval peasant might have never even seen the villagers who live just the other side of the mountain, the residents of modern cities will move through culturally complex and breathtakingly diverse habitats. More than 700 different languages are spoken in New York, a collocated microcosms of humanity that would have been unimaginable even a generation ago (Perlin, 2024). Societal change impacts our sense of self, and even our core biology. According to geneticists, the invention of the bicycle, with its rudimentary expansion of our roaming range, on its own triggered a cascade of human genetic diversification (Vince, 2013).

New technologies and social change are frequently heralded as 'mind-boggling'. For some, this is not just a pithy remark. It is a theory of psychology. Change is feared as an epidemiological risk factor, a precursor to mental illness. Modernity is seen as something that can scramble human psyches. As we saw in Part II, the various transformations of the nineteenth century—the arrival of steam power, the

© The Author(s), under Exclusive license to Springer Nature Switzerland AG 2025   **165**
B. M. Hughes, *Psychology's Quiet Conservatism*,
https://doi.org/10.1007/978-3-032-07724-0_22

availability of daily newspapers, an uptick in 'the mental activity of women', and so on (Beard, 1881)—were believed to have precipitated no less than an epidemic of mental illness. And as we discussed in Part III, socioeconomic shifts that saw women entering what were previously male workplaces were blamed for engendering a host of psychogenic illnesses in those women, ranging from migraine to multiple sclerosis.

Throughout history, it was mostly the poor and the powerless who were said to suffer change-related mental derangement. The privileged saw their own behaviour as representing a timeless psychological normalcy. While we might laugh at the historical scaremongering that warned of dangers such as long-range radio (Robinson, 2016), the typewriter (Stephens, 1998), or pencils that had rubber erasers attached to them (Petroski, 1990), it is worth remembering that the same doomsday narratives also drove protectionist moral panics in response to various political transformations, including the abolition of slavery (Deutsch, 1944) and the granting of votes to women (Willingham, 2021).

In our modern world, perhaps the most commonly cited examples of modernity-related decline involve social media, digital communications, and—because of their ubiquity—portable devices capable of connecting us within seconds to these platforms, primarily smartphones. According to some very esteemed psychologists, the arrival of these technologies has coincided with a sharp decline in the population's well-being. As young people are especially likely to be digital natives, their well-being in particular is said to have been adversely affected.

In 2011, the esteemed British neuroscientist Susan Greenfield claimed that digital technologies had caused young people's brains to 'adapt' to the changing environment, and blamed internet use for causing 'an increase in people with autistic spectrum disorders' (Swain, 2011). When critics pointed out that there was no actual evidence for such a causal link, Greenfield responded by suggesting that simple correlations were enough for her: 'I point to the increase in autism and I point to internet use,' she explained. 'That is all' (McVeigh, 2011). Some years later, Greenfield wrote a book on the matter, in which she cited research published in 2004 to support her assertions about the effects of Facebook and Twitter. The only problem was that

neither platform existed when the studies she mentioned were conducted. After a journalist put this to her in an interview, she pivoted
her argument so that it encompassed more than just social media. She
suggested that 'the existence of email could have had a similar effect'
(Chivers, 2014), now apparently claiming that *email* causes autism too.
Greenfield gained a first-class honours degree in experimental psychology from Oxford University, as well as a doctorate in brain physiology,
before pursuing a renowned academic research career spanning nearly
half a century. Her exploits as a scientist saw her receive many academic
accolades and an appointment to the upper house of the UK's parliament. And yet, despite her high standing, her claims linking social media
use to autism have never been substantiated. And Greenfield has never
submitted any corroborative data of her own for peer-review.

The idea that teenagers risk brain-altering damage from their smartphones and social media coheres with a wider conservative trope of
youth vulnerability and fragility. It fits a stereotype of the young as
displaying timidity, malleability, anger, aberrant attitudes, and styles of
behaviour explicable only with reference to mental pathology. The
manifestations of their mental waywardness are said to encompass not
only the surliness and ennui of lonerism, but also a repertoire of 'acting
out' against their elders, including a general rejection and questioning of
traditional cultural norms. Supposedly encouraged by the 'woke' adults
in their midsts—chiefly their liberal educators and parents—the 'youth
of today' are characterised as overweening and demanding, incapable of
surviving the rigours of daily life without protection from special measures or positive discrimination.

And yet all this coddling is said to be futile. According to a number of
significant thought-leaders in modern psychology (and echoing
Greenfield's own controversial hot takes), the many dysfunctions of
modernity have truly damaged this generation. Social media, smartphones, and excessive coddling, it is argued, have warped teens and
adolescents with wokeism and have doomed them to be triggered by life
itself. The fact that the young are inherently less likely to be socially
conservative is, of course, entirely coincidental.

The conservative critique has thus been legitimised as a popular
assumption: that young people's dissent and rebelliousness relates not

to any valid concerns about societal or global injustice, but instead to literal *damage*—both to their brains and to their psyches—wrought by digital technologies and the distorted worldviews they lead to (including, for example, their facilitation of online radicalisation). A generation of snowflakes has been produced, it is said, and some of psychology's most esteemed researchers claim to have the evidence.

Phone: A Friend

According to the conservative argument, social media saturation, rampant smartphone use, and overprotective coddling by woke parents and educators have combined to produce an 'anxious generation'—a cohort of young people afflicted with mass mental decline resulting from the virtual 'rewiring' of their brains. While this has been something of a recurring trope, it was perhaps most explicitly proclaimed in *The Anxious Generation*, a bestselling book written by prominent US social psychologist Jonathan Haidt (Haidt, 2024a). The book's subtitle—*How a Great Rewiring of Childhood is Causing an Epidemic of Mental Illness*—sets out Haidt's thesis quite succinctly. The very concept of 'an epidemic of mental illness' has historical provenance, echoing almost precisely the language used in George Beard's nineteenth-century bestseller, *American Nervousness: Its Causes and Consequences* (Beard, 1881). Ultimately, as we noted above, the idea that the brains of the masses are addled by modernity is itself not particularly a modern one.

Like Beard's earlier volume, Haidt's book was a massive popular hit, striking a chord with concerned readers not just in the United States, where it spent months on the *New York Times* nonfiction bestsellers list, but also in Europe. Haidt painted a picture of young people enslaved by addictive internet content that had displaced generations-long habits of in-person socialisation and play. Beholden to their smartphones, Haidt argued, young people were now spending 'far less time playing with, talking to, touching, or even making eye contact with their friends and families' (Haidt, 2024b). Social media platforms, he said, were exposing young people to unreasonable life expectations and, especially for young women, to shattering comparisons with idealised body images and beauty standards. Critically, Haidt tied youth immersion in smartphone use to upward shifts in mental ill-health. According to Haidt, the invention of the smartphone had coincided with an extreme mental health crisis in adolescents, characterised by sharp increases in teenage anxiety, depression, self-harm, and suicide. Unlike Greenfield, Haidt was not merely juxtaposing smartphone use with psychiatric statistics and saying 'that is all.' He was explicitly arguing that the former had directly contributed to—i.e., helped cause—the latter.

However, Haidt's causal claims remained rooted in what fundamentally were correlations, and, perhaps worse, in simplistic and at times naive interpretations of basic data trends. It is true that Haidt was able to present statistics showing increases in diagnoses of mental illness in the years following the invention of smartphones. But of the 476 studies cited in *The Anxious Generation*, only twenty-two presented data on either heavy social media use or mental health in adolescents. *None* contained data on both (Brown, 2024). The case being made was almost entirely circumstantial.

This means that *other* circumstances might well explain the outcomes that Haidt described. For example, an important reason why diagnosis patterns change over time is that the methods used to *produce* those diagnoses are frequently updated and revised. In 2011, a mere four years after Apple launched the iPhone, the US government introduced new guidelines for primary care physicians, requiring that female teenage patients be screened for depression at least once a year. This bureaucratic initiative had the immediate effect of significantly increasing the rates of depression diagnosed in young women (Wallace-Wells, 2024). Shortly after, a similar measure was introduced that required hospitals not just to treat injuries, but also to record the *causes* of injuries in patients' medical files. That change alone nearly doubled the number of people recorded as having engaged in self-harm (Wallace-Wells, 2024). And in 2013, the American Psychiatric Association updated the standard method for diagnosing anxiety in its Diagnostic and Statistical Manual (DSM). In the old manual, a diagnosis of anxiety required that a patient be aware that they had a problem with excessive nervousness. However, the update removed this self-awareness requirement, meaning that a person could now receive a diagnosis of clinical anxiety even if they felt their anxiety levels were proportional to the challenges they faced in life (Hughes, 2023). By significantly broadening the inclusion criteria, this change again immediately increased the number of people who received such a diagnosis. Overall, several shifts in diagnostic practice in the period following the introduction of the smartphone could substantially account for the statistical trends flagged by Haidt as evidence of internet-related harm.

Another oversight in Haidt's argument is its US-centrism. His various correlations are vividly contradicted by international comparisons. Teen suicide rates have risen in the US since smartphones became available, but in countries such as Austria, Belgium, France, Greece, Norway, Poland, Spain, and the UK, teen suicide rates have either plateaued or declined during the same period (Roser, 2023). Likewise, the emergence of smartphone use has coincided with rising levels of teen self-harm in the US, but in Denmark, which has one of the highest per-capita rates of smartphone use in the world, teen self-harm has declined sharply over the past fifteen years. In fact, self-harm rates among Danish teens have nearly *halved* since the smartphone was first released (Steeg et al., 2020). And in South Korea, another country where smartphone penetration is particularly high, depression rates for teenagers fell by more than a third in the decade after smartphones were introduced (Kim et al., 2020).

Some US surveys suggest that American teens report less happiness and well-being than they did before they had access to smartphones; however, a plethora of international studies have shown that rates of happiness and life satisfaction among adolescents around the world have either remained static, or have increased, over the past decade-and-a-half (Wallace-Wells, 2024). And we should remember that all these international trends have persisted despite those updates to diagnostic procedures that broadened inclusion criteria and made many forms of mental ill-health *more* likely to be diagnosed than ever before.

Haidt's case draws heavily from cross-sectional, correlational, and observational data, which is inherently limited in its explanatory scope. Many researchers have sought to overcome these problems by conducting extensive longitudinal studies that focus on the sequence of events involved. Large-scale studies are consistent in revealing that teens who have smartphones have *better* well-being than those who do *not* have smartphones—they report greater personal happiness, have higher self-esteem, engage in more physical exercise, spend more (in-person) time with their friends, are less depressed, and are less likely to be cyberbullied than their non-smartphone-owning peers (Martin et al., 2025). Such outcomes are not secondary to socioeconomic factors; in fact, kids from low-income homes tend to be *more* likely to own their own smartphones than those from wealthier backgrounds (Martin et al., 2025).

Overall, longitudinal findings consistently suggest that heavy smart-phone or social media use is just not a precursor to mental illness—in fact, causality emerges *in the other direction* (Odgers, 2024). In other words, difficulties with mental health can themselves lead a young person to increase their engagement with social media (Heffer et al., 2019). When a young person experiences a mental health difficulty, they are likely to turn to their smartphones and social media as part of their attempt to cope with what is happening to them. The unfettered flow of online knowledge has undoubtedly contributed to greater mental health awareness across the world, as well as to destigmatisation. The internet can provide teens with forms of social support that would have been unavailable to their parents, as well as access to information that assists them in seeking and securing appropriate help.

Once again it is useful to reflect on the distinction between the *prevalence* of mental ill-health and its *diagnosis*. Rates of diagnosis might change for reasons unrelated to underlying rates of prevalence. Part of the reason many young people come to get diagnosed with anxiety or depression in the first place is because they had the chance to be informed—online—about the nature of their own mental health. They seek assistance from healthcare providers, which benefits their lives, precisely because of what they have learned from the internet and social media. So yes, for these young people, the end result would be a link between internet use, mental health status, and service contact. However, not one that supports Haidt's theory—but one that directly contradicts it.

23

Depoliticising Youth Anxiety

The reality, of course, is that datasets focusing only on smartphone use and mental health are inherently reductionist: they predispose a reader to jump to simplistic causal conclusions because they artificially focus on a small number of possible factors, with little basis to infer direct links between them. Anxiety, depression, self-harm, and suicide are highly complex problems with multifactorial causal contexts. Some of these contexts relate to societal, economic, political, and cultural conditions; others relate to biological aspects, such as genetics. Many of the risk factors are intergenerational. While research has consistently failed to link internet use to mental health problems, it has time and again found many other variables to be statistically significant predictors.

Financial hardship, structural discrimination, economic inequalities, homelessness, parental mental health problems, having a sibling with a disability, exposure to violence, witnessing racism, being targeted by sexism, or being subjected to child abuse and neglect are just some of the variables to have been identified as solid causes of mental ill-health in young people. The researchers who have delineated this nexus of adverse childhood experiences often encapsulate them under the heading of 'toxic stress'—namely, factors that lead to stress-response activation in

173
B. M. Hughes, *Psychology's Quiet Conservatism*,
https://doi.org/10.1007/978-3-032-07724-0_23

a way that persists beyond what is tolerable. Their work has led to the emergence of what specialists refer to as an 'ecobiodevelopmental' theoretical framework (Shonkoff et al., 2012). In short, there is a lot of good science on this issue. Just very little that has been produced by psychologists.

Psychologists' focus on individuocentrism steers them away from thinking about economic, societal, political, or cultural risk. Their insistence on quantification inclines them to prioritise concepts they can visualise and measure over ones that are more amorphous but possibly also more important. And their tendency toward reductionism encourages them to select concise sets of candidate variables that they then analyse in isolation, and to implicitly (but irrationally) dismiss as irrelevant any variables they have *not* chosen.

Haidt, for example, has claimed that while the invention of smartphones in 2007 unleashed an epidemic of mental ill-health, the Great Recession—the sharp downturn in the world's economies that occurred between December 2007 and June 2009—had no lasting effect on people's well-being. He defends this hypothesis by pointing out that while post-Recession unemployment rates recovered during the 2010s, the mental health of young people (in the United States) supposedly did not. According to Haidt, the absence of such a correlation disproves his critics' claims that young people's mental health is impacted by economic conditions. However, there are at least two problems with this.

Firstly, a focus on headline unemployment rates is meaningless when you consider that not everyone is affected equally by recessions. What is missing from national unemployment statistics—which tell us the state of the *total* population—are the particularly harsh experiences felt by the *most vulnerable subsets* of that population. People who were economically marginalised prior to the Great Recession saw their already meagre prospects damaged even further—and in many cases irredeemably—by the downward economic spiral that started in December 2007 (Barr et al., 2012). The rate of overall unemployment tells us little about how unemployed people are faring. Falsely presuming that a single, unitary, summarising-the-entire-population statistic can help reveal how surrounding conditions affect each *individual* within the population is an example of a reasoning error known as the 'ecological fallacy'.

Methodological reductionism makes psychologists especially prone to precisely this type of erroneous extrapolation.

Secondly, there is much more to an economic recession than unemployment. Millions of people who kept their jobs during the Great Recession were still impacted negatively by it. They were left facing a lifetime of flattened income, rising living costs, ever-mounting indebtedness, precarious housing arrangements, reduced prospects, and many other economic legacies that are toxic to mental health. We know too that recessions serve to increase social inequality—the gap between rich and poor—a metric that itself is known to damage people's mental well-being (Yu, 2018). This is not even to mention the widespread cutbacks to mental health services that were implemented during the recession in the name of economic austerity, depriving those who present with mental health problems of supports that might help them (Odgers, 2024). And surely even rich people, young or old, can become distressed by witnessing the adverse impacts of recession on those around them, such as increasing rates of poverty and homelessness in their communities. Indeed, the presumption that a person's well-being will remain intact so long as their *own* income is protected is just another example of capitalist realism in psychology. Economies are famously multifaceted; focusing on unemployment statistics is an incomplete way to judge their effects. A global recession of the scale of that seen in 2007–2009 will harm human well-being in a multitude of ways.

And aside from economic decline—but not always unrelated to it—young people can also no doubt be made anxious by the range of interwoven existential threats that they see vividly in their everyday lives. These include the demise of democracy, the rise of the Far Right, and the emboldening of social prejudices, as well as geopolitical instability, televised warfare, and the displacement of refugees. There is also food insecurity, infectious disease, and the climate crisis, each of which raises dire implications for the future habitability of the very planet on which their lives depend. In the US, where Haidt formulated his worldview, you can add to that list an unprecedented opioid crisis and apparently out-of-control levels of gun crime. Simply put, the young people Haidt talks about have very many reasons to be more anxious than they used to be.

Claiming that mental health is being harmed by smartphones and social media (or, for that matter, by wokeness) rather than by obvious social and ecological denigration of the human habitat serves little tangible purpose other than to draw attention away from the political system and its many masters. Banning TikTok or putting kids' phones in locked boxes during the school day will do little to improve young people's well-being—but it will certainly help to perpetuate the status quo, by distracting the masses.

24

Biological Reductionism Revisited

Core to concerns regarding the dangers of modernity is a set of assumptions about what is 'natural' for human beings to have to deal with, compared to what is 'unnatural' and so challenging to their inherent coping capacities. Whether stated or merely implied, these assumptions include the view that human minds (and bodies) have, over time, become adapted to particular living environments and social conditions—chiefly those that have predominated throughout our ancestry. In short, it is assumed that humankind has been shaped by its past. This basic assumption underpins the socially conservative worldview: given that human beings are better suited to coping with the environments and conditions of the past, it follows that *environments and conditions that resemble those of the past are better suited to human beings.*

From this it can be argued that modernity itself presents a range of species-wide threats to human welfare. High-fat diets can be blamed for the modern proliferation of heart disease and diabetes, because our Stone Age appetites were adapted for food-scarce environments and not for the ultra-calorific societies in which we now live. Sedentary lifestyles can be implicated as adding to the obesity epidemic by suppressing our ancient human impulses for physical activity. Work practices such as night shifts

B. M. Hughes, *Psychology's Quiet Conservatism*,
https://doi.org/10.1007/978-3-032-07724-0_24

can be presented as contributing to widespread ill-health by forcing human beings into unnatural states of night-time wakefulness. Women's participation in industry or higher education can be framed as fundamentally challenging to the female body's evolved capacities, which were adapted for lives devoted to motherhood and childcare. Gender fluidity and homosexuality can be cast as interrupting the heretofore unbroken chain of human reproduction that connects every living modern human being to their ancient evolutionary ancestors. And smartphones and the internet (or even email) can be considered to be existential threats to humanity by portraying them as disrupting our otherwise natural patterns of in-person socialisation.

'The old ways are the best' is not just traditionalism for its own sake, it is an argument premised on (scientific) assumptions regarding what is believed to be natural for human beings. By asserting that human behaviour is fundamentally rooted in brain function, physiology, and genetics, psychology's biological reductionism lends scientific credibility to this conservative worldview. It also serves to *encourage* it, by conveying the message that naturally occurring human attributes represent 'fitness', are 'adaptive', and therefore must be 'good'. Where nurture is altruistic and thus laudable, nature is depicted as intrinsically moral and therefore unassailable.

What psychology students are taught to call the 'nature–nurture debate' presents a fundamental political cleavage as scientific discourse. Those who espouse an entirely 'nature'-based position believe that human behaviour is determined by essential biological forces that are innate and inherited. Those who espouse an entirely 'nurture'-based position believe the opposite: that human behaviour is determined by external forces that influence a person's development through perception, experience, and reinforcement. The former posits that human beings are what they are and should therefore be held personally accountable for all that they do, and that intervening to improve matters that are fundamentally determined by nature is futile and consequently wasteful. By contrast, the latter position implies that people should *not* be regarded as irretrievably immutable, but should be seen as the products of their environments—and thus that policies aimed at improving matters by modifying those environments are always worth considering. In short, the 'nature' position is the theoretical basis of social conservatism, because it espouses themes such as personal responsibility and fixed human hierarchies; whereas the 'nurture' position is the theoretical basis of social liberalism, because it promotes such principles as social responsibility, social mobility, and social intervention—effectively, what historically have been cast as hyper-liberal 'big government' and 'social justice' notions.

Of course, psychology students are also taught that *neither* 'nature' *nor* 'nurture' is ever solely decisive, and that there is always a combination of both in play. However, this implication of neutrality is misleading. Unless the universe is somehow cosmically arranged into an arithmetically perfect balance, it is impossible for the forces of 'nature' and 'nurture' to ever be exactly equal, as if cancelling each other out in importance. To state that both are in play involves acknowledging that one of them will inevitably have a greater influence than the other. The vagueness of the 'bit-of-both' stance simply enables confirmation bias to be unleashed. Theorists swayed by an unconscious bias remain free to favour the particular force that more closely reflects their politics.

In reality, despite the fact that psychology is so regularly accused of being liberally biased, the apparent consensus among psychologists is that 'nurture' explanations are subordinate to 'nature' ones, not the other way around. Human beings have naturally endowed brains, after all, and it is the capacities of the natural brain that set the parameters for human behaviour. All the so-called 'nurture' theories are inherently nature-bound. Social cognition, threat vigilance, positive reinforcement, developmental stages, group influence, altruism, and so on, are each framed as being subject to a person's genetic blueprint, hormone levels, or instinctive (i.e., innate) responsivity to social incentives. The centrality of brains to psychological function is undisputed. Nobody doubts that brain damage is immediately destructive to cognition. But the idea that a human child is some kind of *tabula rasa* whose entire psychological life is decided by social imprinting is regarded as a naive historical hypothesis, useful only as the basis for thought experiments and classroom discussions. Psychology's 'bit-of-both' resolution demands that, while both are relevant, the 'nature' bit will always delimit how the 'nurture' bit can function.

Therefore, in the context of confirmation bias, it is important to note that the apparent bias within the field of psychology is toward confirming, and thus reifying, the primacy of 'nature' explanations. Notwithstanding psychologists' strong focus on developing evidence-based interventions, their field's accounts of human behaviour consistently sanctify the socially conservative end of the nature–nurture spectrum. Conservative arguments are typically underpinned by biological rationales. Accordingly, they are easily bolstered by drawing on the psychology curriculum, with all of its biological reductions.

A sobering observation here is that drawing extensively from psychology does not appear to have infused conservative arguments with any conspicuous rigour or consistency. Psychology's knowledge base, theories, and methods of explanation seem vague enough to support a variety of contradictions, resulting in a discourse that is peppered with ironies. One of these ironies relates to the conservative attitude toward threats. On the one hand, social change, modern technology, and left-leaning wokism are all depicted as posing ongoing threats to humankind, because they destabilise the homeostasis of civilisation. Immediate action is said to be required on the basis that threats are inherently bad. But on the other hand, whenever people on the left complain about things that threaten *them*, they are ridiculed for being snowflakes. They are advised that threats are an unavoidable part of life and that navigating the school of hard knocks is an important part of being a successful adult. It toughens you up. In this context, threats are portrayed as inherently *good*, on the basis that facing them head on helps people to build resilience and robustness. The fact that threats are simultaneously bad *and* good reflects the central ambiguity of conservative concerns about coddling: the belief that young people supposedly need protection ... from the plight of being over-protected!

A second irony arises from the realpolitik of the culture wars. Conservatives often rely on psychobiological concepts that are deeply rooted in evolutionary theory—the gist being that modern living brings societal changes that fly in the face of nature by subverting the effects of natural selection. These evolutionary ideas can be broad or specific. In broad terms, conservatives regularly warn against upsetting the finely poised balance that has emerged (naturally) in our societies, even if those societies are characterised by poverty, division, and hardship. Suffering, they argue, is endemic in nature, and so should be seen as unavoidable. ('There Is No Alternative.') The specific arguments include essentialist claims about the behaviours and capacities of different population sub-groups. Evolution is the premise whenever conservative commentators imply that certain groups in the population are prone—i.e., genetically endowed—toward criminality, or that some groups (usually the same ones) are predestined—i.e., genetically—to fare less well in the education system. Conservatives also invoke evolutionary principles when objecting to social safety nets, which are accused of undermining nature's inherent 'survival-of-the-fittest' dynamic. Conservative defences of traditional gender roles and behaviours (up to and including male violence) are also bound up in evolutionary theory: their implication is that 'naturally occurring' behaviours should be accepted as normal, and indeed as *healthy*, because they represent, *ipso facto*, an organically determined balance of human attributes.

At the extreme end of conservatism, evolutionary principles are inter-twined with the dogma of eugenics. Evolution is the implied basis for objecting to inward migration, especially where it contributes to greater ethnic diversity. A common eugenicist conspiracy theory warns of a supposed worldwide effort to replace White people with non-White people by engineering migration flows in ways that tilt the balance of natural selection. Overall, a whole raft of conservative fears are framed in terms of their supposed implications for 'future generations'—itself an allusion to the human gene pool as conceived of in evolutionary terms.

The nub of the irony is that all these complaints invoke evolutionary principles and yet all are frequently favoured by hardline religious conservatives whose extremist doctrinal leanings predispose them simul-taneously to deny, as sacrilege, the theory of evolution itself. A significant subset of conservative voters are motivated by extreme, if not fundamentalist, religiosity. It is not unheard of for conservative advocacy groups to warn of looming (White) racial demise due to gene pool depletion, while simultaneously lobbying for a ban on the teaching of evolution in schools on the grounds that it contradicts scripture and so cannot be true. In short, Darwinist principles are frequently pushed hardest by people who otherwise decry Darwinism.

Part V

The Coddling of Conservative Minds

25

How Psychology Reinforces (and Thus Perpetuates) Social Conservatism

The problem with biological reductionism in psychology is not that evolutionary theory is flawed. It is that there is more to the human experience than biology. As such, biological reductionism doesn't just furnish psychology with incomplete explanations, it helps to implant a socially conservative bias at the heart of psychologists' *modus operandi*. Even in the twenty-first century, psychologists maintain an allegiance to individuocentrism, reductionism, and styles of quantification that distil human minds down to simplistic stereotypes, thereby accentuating the conservative leaning of their field. They persist in portraying human beings in capitalist realist terms, as self-interested, competitive, and materialistic. And they produce explanations of how humans think, feel, behave, and prosper that omit important political context. Far from being liberally biased, psychological discourse frames humanity as a species of sovereign individuals who are naturally primed to protect their own interests, who are endangered by deviations from traditional norms, and who are encouraged to seek happiness within adversity by regulating their emotions in ways that discourage them from panicking too much over the state of the world.

© The Author(s), under exclusive license to Springer Nature Switzerland AG 2025   **185**
B. M. Hughes, *Psychology's Quiet Conservatism*,
https://doi.org/10.1007/978-3-032-07724-0_25

The mechanics of psychology's systems of knowledge-generation are geared to reject accounts of human predicaments that emphasise power structures or political forces. Instead they promote a style of explaining that sees destiny as determined primarily by one's own decisions. In all these ways, psychology is effectively the *antithesis* of progressivism. It validates and thus preserves the status quo, with all of its injustices, by framing it as apolitically 'natural'.

Time and again, when presented with controversy, psychology's curriculum defaults to conservative resolutions. Let us now consider eight examples of how this happens.

26

Example #1: By Standing Up Against Safetyism

It is no coincidence that the prequel to Haidt's *Anxious Generation* was another bestselling treatise on how modern mores sabotage the welfare of today's youth. In 2018, Haidt released *The Coddling of the American Mind: How Good Intentions and Bad Ideas Are Setting Up a Generation for Failure*, a book co-authored with lawyer and activist Greg Lukianoff (Lukianoff & Haidt, 2018). In it, Haidt and Lukianoff argued that the youth of today are being overprotected by woke educators and parents. Specifically they took aim at college campuses. They decried universities' use of 'trigger warnings' and 'safe spaces', which they said undermine free speech by pandering to unwarranted sensitivities. They also rebuked several social science concepts typically raised in discussions of power abuses, including 'microaggressions', 'intersectionality', and 'identity'. And they poured scorn on the suggestion that people should feel entitled to 'emotional safety', a cause they labelled as *safetyism*. For Haidt and Lukianoff, allowing safetyism to prevail was akin to conceding to 'coddling', the particular evil they accused of setting up the younger generation 'for failure'. But all was not lost, because Haidt and Lukianoff had some solutions. They recommended that all this danger could be dealt with if universities signed a pledge to ban trigger warnings and safe spaces, and if young people were trained, using cognitive behaviour therapy, to stop worrying.

But although conservatives freely mock the use of trigger warnings in higher education, seldom do they acknowledge the subtleties involved. In reality, trigger warnings originated in online communities as a way to advise readers that content on a particular website might include references to rape, child abuse, domestic violence, or other subject matter that a reader who themselves was victimised by these crimes would wish to know about in advance. They are, in fact, rarely ever used in universities. When they are, it is to signal support for students who may end up encountering traumatising material, rather than to prevent them from actually doing so (Vivian, 2023). In other words, trigger warnings do not prevent students from being taught about rape—they simply inform students *who themselves might have been raped* that rape is on the syllabus. It seems difficult to see quite what is so inhumanely coddling about all this.

Likewise, when conservatives ridicule the idea of 'safe spaces' on university campuses, they ignore the history of this term and its genesis in protests by Black and other historically disenfranchised students against on-campus racial harassment in the United States. The trope of safe spaces became concretised in popular discourse following demonstrations at the University of Missouri in 2015 and 2016. The protests occurred against the backdrop of centuries of racial segregation in the state of Missouri, in which the university itself had historically been implicated (Vivian, 2023). Students sought safe spaces in response to a series of hate-speech incidents at the university that included the smearing of a swastika, using human excrement, onto a Black student's dormitory wall. Contrary to conservative dismissals, the dimensions of this issue extend far beyond its reductionist stereotype, which frames safe spaces as crèche-type facilities provided by woke universities to shield spoiled students from viewpoints they might not agree with.

The fact that Haidt and Lukianoff's treatise was littered with socially conservative talking points might have caused liberally-minded psychologists to quickly dismiss it. The authors' own reputations would have compounded the effect. Haidt is a well-known critic of the supposed liberal bias in psychology and has frequently complained that his field is biased against social conservatives (e.g., Haidt, 2016). So, while ostensibly couched with reference to empirical datasets, any arguments he mounts about coddling (or smartphones) should rightly be evaluated in the context of his broader political views. Lukianoff, for his part, is also a high-profile conservative commentator, whose own previous writings have railed against the supposed censorship of conservatives by universities (e.g., Lukianoff, 2014). For nearly two decades he has served as president of FIRE, the American anti-woke lobby group we discussed in Part I.

But despite the presentational red flags, Haidt and Lukianoff's argument simply reflected a discourse that has become conventional in psychology. Their points mirrored those put forward as standard, over the past hundred years, throughout psychology's orthodox textbooks and journals.

Like most mainstream psychologists, Haidt and Lukianoff framed their position in terms of individuocentrism. They focused on personal traits and behaviours while largely ignoring social, political, economic, or cultural processes. In their view, personal traumas should be dealt with by cultivating individual robustness and not by embarking on social action; people should care for themselves, rather than expecting others to support them with trigger warnings, safe spaces, and the like. Haidt and Lukianoff's analysis also echoed mainstream psychology's reductionism—it relied on monofactorial explanations, in which problems have single causes (such as coddling) rather than systemic ones (such as structural discrimination). And for good measure, Haidt and Lukianoff also advocated mainstream psychology's capitalist realist portrayal of people as competitive consumers—the notion that healthy human beings should always be seeking to better themselves as individuals by striving for greater self-esteem and personal growth, rather than, say, by espousing public citizenship, collectivism, or civic action.

A further reflection of mainstream psychology in Haidt and Lukianoff's treatise was the belief that if a core problem is individual, then its solution should be personalised—specifically, some form of psychological intervention (in this case, CBT). Not only does this further resonate with individuocentrism, it also reflects a commonly seen *psychologists-to-the-rescue* narrative, in which psychologists portray themselves as the main characters in a heroic problem-solving drama. As we saw earlier, this tendency toward professional self-aggrandisement is partly why psychologists tend to overestimate their ability to explain the causes of mystery illnesses. (We will return to the broader issue of exaggeration in psychology in Part VI.)

Overall, conservative complaints about safetyism, such as those of Haidt and Lukianoff, draw on—and mirror—the basic worldview of psychology. They share the field's assumption of individuocentrism, its habit of reductionism, and its embrace of capitalist realism, framing all human beings as simplistically self-interested agents. In reality, societal problems are multifaceted, somewhat abstract, and maddeningly recursive; they do not easily lend themselves to sparsely phrased hypotheses or statistical summary. Despite this, psychology's approaches drive discussion toward totemic ideas that are easy to visualise and debate, and which cohere to conservative analyses rather than reflect their true complexity.

27

Example #2: By Pathologising Dissent

The notion that people are being 'coddled' is itself patronising and demeaning. As a response to claims that too much coddling is damaging their mental health, offering CBT to toughen up today's youth is little more than establishment gaslighting. It continues psychology's tradition of training people to eschew their emotional reactions to societal problems and to reframe their perspectives so as to not see them as problems per se.

The conservative right has a long history of downplaying dissent by framing it as pathology. Recall that in Part II we saw how nineteenth-century doctors claimed enslaved people who attempted to escape captivity were in fact suffering from a psychiatric condition called *drapetomania* (Guthrie, 2004). For good measure, the physician who named that condition argued that its main cause was the unwarranted sympathy shown by slave 'masters' who befriended enslaved people and started 'treating them as equals' (Cartwright, 1851; p. 708). In other words, drapetomania was blamed on excessive coddling by woke plantation owners. Nineteenth-century doctors viewed slavery as acceptable and normal; they believed that only people who suffered from a mental illness would protest against it.

B. M. Hughes, *Psychology's Quiet Conservatism*,
https://doi.org/10.1007/978-3-032-07724-0_27

A century later, American psychiatrists drew on precisely the same reasoning to coin the label *protest psychosis*, used to imply that Black people angry at the denial of their civil rights were in fact displaying symptoms of schizophrenia (Metzl, 2010). This was during the 1960s, a period when the American Psychological Association was encouraging its members to remain silent on the matter of racial equality, on the grounds that it would be inappropriately political for a professional psychologist to object to society's norm of discrimination (DeAngelis & Andoh, 2022).

Decades on, such examples might feel outdated. However, the pathologising of dissent remains common in contemporary society. Psychologists frequently frame mental illness as arising from the disjuncture between a person's own functioning and the expectations of the society around them. Conservatives adopt a similar narrative when depicting political dissent: the established order of the day is presented as the 'norm', and protesters are portrayed as 'challenging', or deviating from, this norm. In both instances norm violation is problematised. But, by its very nature, protest demands the breaking of social rules. Stripped of their context, acts of protest will always seem erratic. When the basis of a protest is left unclear—either deliberately or in the name of journalistic neutrality—vivid dissent can easily appear frightening, as though society itself has begun to disintegrate.

The inherent ambiguities make it easy for conservatives to dismiss mass protests as irrational, impulsive, immature, violent, or unstable. In 2020, when anti-racism protesters removed a statue of a seventeenth-century slave trader from a public square in the British city of Bristol, the right-wing press portrayed them as part of a 'feverish crowd' whose actions displayed 'mindlessness' (Kirby et al., 2022). Right-wing commentators frequently dismiss climate activists as 'lunatics' (Foster, 2022), 'mindless louts' (Oakeshott, 2024), 'narcissists' (Malone, 2024), or 'nutcases' (Davies, 2023), and condemn environmental protest activities for being 'demented' or 'deranged' (Pankratz, 2024). Students participating in demonstrations related to the Palestine–Israel conflict have been derided by conservatives as 'mad brats' (Strimpel, 2024), as 'narcissists' and 'psychopaths' gripped by 'fantasies' (Peterson, 2023), and as being 'detached from reality' as a result of 'mass hysteria' (West, 2024). In 2024, after such anti-war protests erupted in university campuses around the world, one medical commentator went so far as to claim that the students were exhibiting symptoms of 'mass psychogenic illness' (Christakis, 2024), as though their emotional reaction to what they saw as their colleges' complicity in the slaughter of innocent civilians was merely a moment of youthful madness.

Sometimes the narratives seem deliberately geared toward psychological distancing. Protest groups are given simple titles that come to supplant any detailed discussion of their cause. This is the very epitome of reductionism. By recasting diverse movements as homogenous out-groups, blobbish nicknames—such as 'BLM', 'Antifa', 'the Radical Left', 'the wokerati', 'Remoaners', 'Zapatistas', 'eco-yobs', 'anarchists', 'SJWs', or 'gender zealots'—are deployed to heighten protestors' apparent separateness from ordinary people. This deliberate 'exoticising of resistance' (Theodossopoulos, 2014) implies that dissenters occupy an unnatural social space, and that the act of dissent itself constitutes a departure into some esoteric state of thinking. The caricaturing of resistance exaggerates the differences between protest groups and mainstream society, and minimises the impression of common concerns that might unite the population.

The point here is that possessing a different opinion from others should not be seen as a sign of mental disease. And yet psychology's individuocentrism, reductionism, and capitalist realist narratives all lead to precisely this implication. Dissent is to be viewed as a person-level behaviour, not a consequence of societal conditions. Causes are to be simplified, groups stereotyped, and quick-fix solutions proposed as remedies. And we are to see all human beings (or at least the right-minded ones) as being motivated by the demands of market capitalism. Psychology's conventional assumptions inherently require us to view dissent as potentially or actually maladaptive: if compatibility with social norms is a sign of health, then the active rejection of those norms must amount to extremism. According to psychology, dissent is a sign of chaos, rather than of the need for change.

As such, whenever psychologists discuss dissent in terms of identity, groupthink, impulsivity, obedience, frustration–aggression, deindividuation, social influence, or social contagion, they do so largely while presuming the status quo to be desirable and protest to be threatening. In other words, they draw on socially conservative frames of reference.

28

Example #3: By Labelling Deviance

Not only do the norms of psychology pathologise political dissent, they also allow *genuine* pathologies to be dismissed, downplayed, or distorted by political bias. As discussed above, the choice of which behaviours count as deviant—and which don't—resides entirely at the discretion of psychologists. This ensures that the conventions of clinical psychology evolve in ways that reflect the political worldviews of those who exert influence upon them. Psychology might aspire to be a clinically relevant science, but its methods for classifying mental ill-health are arbitrary and culturally contingent. It defines mental illness selectively, influenced by individuocentrism, reductionism, and capitalist realism. In other words, it does so using processes that lean toward socially conservative depictions.

Not all violations of social norms are classed as pathological. White-collar crimes, such as insider trading or price-fixing, are seldom considered examples of mental illness even though they clearly violate the norms of social behaviour. But other violations are considered differently. Persistently defying requests from authority figures, for example, is liable to be treated as a sign of a conduct disorder. The inconsistencies often hint at sociopolitical undercurrents. For example, in the United

States, children from racial minority communities are significantly more likely to receive a diagnosis of 'oppositional defiant disorder' than children from the majority White population, whereas White children who present *with the same behaviours* are more likely to be diagnosed with 'attention–deficit/hyperactivity disorder' (Fadus et al., 2020). There is no actual racial divergence in the prevalence of these two conditions. It is just that, for whatever reason, clinicians tend to choose the more pejorative-sounding label when a child belongs to an ethnic out-group.

Such clinical open-endedness serves to facilitate, and perhaps encourage, political projection through diagnosis. As we have seen, right-wing commentators frequently employ diagnostic terminology to attack political and cultural out-groups, such as when pathologising political dissent ('protest psychosis' and 'mass psychogenic illness' being just two examples). But it also creates a double-standard—on other occasions, the very same right-wing commentators will *decry* their political opponents for depicting their *own* lives in diagnostic terms.

In 2016, after the singer Lady Gaga explained she had experienced PTSD as a result of being raped as a teenager, a high-profile British conservative television presenter proclaimed that terms such as 'post-traumatic stress' should only ever be used to refer to military personnel and were never appropriate for civilians. 'Enough of this vain-glorious nonsense,' he wrote in a social media post. 'It angers me when celebrities start claiming "PTSD" ... to promote themselves' (Morgan, 2016). In reality, he was entirely wrong on every aspect of his point. PTSD is, in fact, *more* prevalent among civilians than it is among the military. And it is far more common in women than in men precisely *because* women are more likely to be sexually assaulted (Olszewski & Varrasse, 2005).

Conservative commentators are frequently triggered by clinical language. They complain that diagnoses such as Tourette's syndrome and obsessive-compulsive disorder have acquired 'social currency', as though people *want* to be labelled with such terms, perceiving them to be fashionable (Daum, 2022). In some cases, they go so far as to reject the very reality of diagnosed conditions. Many direct particular ire at the concept of 'gender dysphoria', claiming that no such thing exists. In their eyes, the term is a politically-motivated designation, contrived solely to legitimise the provision of gender-affirming care to trans youth against the wishes of objectors. In 2025, a prominent British conservative columnist claimed something similar about dyslexia. 'It's a made up affliction,' he wrote, 'that's become a multi-million pound industry around children who haven't been taught to read' (Hitchens, 2025).

The problem is that all psychology's clinical categories involve an element of interpretation. Any diagnostic label can be said to be constructed on no more than shifting sociocultural sands. Anyone can, for political reasons, dismiss a particular diagnosis as a fad.

Mental health conditions are seldom divorced from social or cultural context, but it is only ever the *person* who is diagnosed as disordered. As such, it is the *person* who must carry the label. The overall effect of this form of individuocentrism is to focus attention on problems that arise within people, rather than in society. The blame is laid on people's negative traits, poor choices, destructive actions, low abilities, or risky proclivities. When toxic environments engender mental anguish, as they often do, the diagnostic architecture of clinical psychology's taxonomy sees only defects within sufferers, not dysfunction within the world.

The defect approach holds individuals responsible for their own difficulties, and so legitimises social marginalisation. It disproportionally affects out-groups. It has long been observed that in many countries, such as the United Kingdom, people with minority ethnic backgrounds are much more likely than persons from majority communities to be detained under mental health legislation (Singh et al., 2007). Several studies have shown that immigrants are far more likely than non-immigrants to be diagnosed with schizophrenia (Schofield et al., 2017), anxiety (Abebe et al., 2014), depression (Malmusi et al., 2017), or eating disorders (Strand et al., 2023)—despite the fact that they are generally *less* likely to seek out mental health services in the first place (Jing et al., 2023). Overdiagnosis of mental health problems in culturally marginalised out-groups not only compounds the many pre-existing social disadvantages that they face, it effectively amounts to a form of racial profiling. It seems very much at odds with claims that the field of psychology is skewed by 'liberal bias'.

Psychologists are viewed as clinicians and their field is widely believed to be a science. So when they declare that unruly ethnic-minority children have a disorder marked by 'defiance' whereas similarly behaved White kids are merely 'hyperactive', the distinction will likely be seen as being informed by both professional experience and empirical evidence. The fact that it might actually reflect unconscious racial bias is typically not appreciated by the general public, or even by psychologists themselves.

29

Example #4: By Stigmatising Negativity

The nuances arising from psychology, politics, dissent, and deviance are often shaped by value judgements regarding the merit of negativity. In short, whether a person's frame of mind is 'positive' or 'negative' is deemed to be core to their mental functioning. The distinction is itself politically valenced. 'Positive' frames of mind are seen as constructive and healthy, while 'negative' ones are said to be toxic and disordered. Positivity is duly portrayed as an intrinsically desirable mental state. Overall, psychology casts 'negativity' as corrosive and discreditable. Positive people are to be embraced, negative ones are to be avoided.

It is culturally commonplace for 'negative' emotional styles to be harshly stereotyped. Those who display them are likely to be cast as 'party poopers', 'buzz killers', or 'vibe wreckers'. But it is also common for persons with serious depressive illness to encounter an extension of this stigma. Several studies have highlighted how sufferers of major depression customarily face discrimination and alienation. One recurring theme in the research is that the stigmatising of clinical depression is intertwined with critics' political worldviews (Thibodeau et al., 2015). In particular, social conservatives are especially likely to hold hostile attitudes towards sufferers of depression. People who endorse conservative ideologies are likelier than their peers to believe that a clinically depressed person is 'unpredictable' or 'dangerous'. In an echo of the 'malingering' narratives we encountered in Part III, conservatives are more likely to believe that a person with depression could simply 'snap out' of their emotional state if they really wanted to (Löve et al., 2019). Conservatives also tend to avoid spending time with depressed people, reporting that they would rather not live near, work with, or befriend a person who had such an illness (DeLuca & Yanos, 2016). For many conservatives, negative emotionality is just inherently *bad*, and not just for those who experience it.

Mainstream psychology has traditionally endorsed the inherent wholesomeness of positivity. It has done so in at least two ways. Firstly, it has expended great effort developing clinical therapies that explicitly aim to extinguish *negativity*, on the clear basis that negative emotions are unwanted and unhealthy. And secondly, it has accommodated the emergence of a new-ish free-standing specialist subfield that explicitly seeks to promote a focus on the positive. This subfield, branded 'Positive Psychology' by its proponents in the late 1990s, aims to ensure that psychology emphasises 'the good life—what it is to be healthy and sane', and to displace psychologists' more orthodox interest in 'studying and treating misery' (Seligman, 2019). Positive Psychology encourages practitioners—and the wider public—to avoid focusing on negative feelings, and to instead spend time fostering positive experiences, attitudes, and traits. A fundamental assumption of Positive Psychology is that our future welfare is determined by the nature of our current anticipation. This means that mental health is to be achieved by replacing hopelessness with a mindset of determined optimism (Duckworth et al., 2005).

But while *extreme* depression can undoubtedly produce unbearable emotional pain, it is less clear that milder sadness is something we should avoid at all costs. The capacity to experience negative emotion is an important protective mechanism. For one thing, it provides us with aversive feedback from damaging experiences, thereby reinforcing us to avoid repeating them. It also signals our distress to people around us, making it more likely that someone will come to our aid. Social cohesion requires us to fully experience our negative emotions so that we can sympathise with others and regulate our behaviour to reduce any harm we might be causing. It is well understood that people who are incapable of negative emotion have problematic personalities; indeed, impairment of negative emotion—such as an inability to feel empathy or an immunity to stress—is one of the main characteristics of psychopaths. The problem is that Positive Psychology's focus on positive emotion downplays the importance of negativity to overall human well-being. Encouraging people to be zealously positive as a means of suppressing distress could be seen as akin to asking them to feign psychopathy in order to remain unperturbed by social injustice.

Positive Psychology has been critiqued many times (Frawley, 2015). Its use of algebraic formulae to analyse 'authentic happiness' has been accused of being pseudoscientific and mathematically incoherent (Pérez-Álvarez, 2013). One of its highest-profile theoretical constructs, the idea that psychologists can compute a person's 'positivity ratio' in order to predict their future well-being, has also been statistically debunked, and condemned as a gullible exercise in 'wishful thinking' (Brown et al., 2013). Positive Psychology's focus on positivity in interpersonal inter-actions has been shown to be a double-edged sword, with several long-itudinal studies revealing that the promotion of optimistic expectations and unconditional trust leaves people vulnerable to manipulation by others (McNulty & Fincham, 2012). And, with no little irony, many of the subfield's own pioneers have been accused of being difficult to deal with in person, and unpleasantly hostile toward criticisms of their ideas (Ehrenreich, 2009).

But more importantly, Positive Psychology operates in subtle ways to reinforce conservative political values. Firstly, it helps to pathologise political dissent by problematising negative feelings—including those that might understandably be provoked by injustice and inequality. When viewed through the lens of Positive Psychology, political protest is just another form of negative thinking. Far better, the subfield recommends, to focus on what is *good* about the world than to dwell on its darker side.

Secondly, Positive Psychology adopts an individualist, reductionist, and somewhat capitalist-realist approach to the idea of happiness. By facilitating the notion that human beings should strive for ongoing optimism, it creates an unrealistic expectation of constant human improvement. People are encouraged to be aspirational consumers when it comes to their mental well-being. They are told that they must always visualise a future that is more positive than the present, an exhortation that sets an impossible standard for happiness (Fernández-Ríos & Novo, 2012). Moreover, by disparaging negative emotion, Positive Psychology encourages the belief that *un*happiness is a type of mental dysfunction, for which therapy—or, at least, the advice of Positive Psychologists—should be prescribed. It thereby perpetuates the medicalisation of emotion and the marketisation of human contentment (van Zyl et al., 2024).

This depiction of happiness as a commodity—central to a good life and achievable, tantalisingly, through individual mental adjustment—is deeply individuocentric. It presents self-fulfilment as a personal enterprise and human welfare as the responsibility of individuals themselves rather of the societies or institutions with which they interact (Yakushko & Blodgett, 2021). It encourages people to consider their every choice and behaviour in instrumental terms, as efforts to be deployed in a never-ending rat race toward optimal contentment. And more prosaically, by normalising expectations for idealised forms of happiness—and implying that human beings are authors of their own fortunes—it creates a lucrative commercial market for self-help books, consultancy appointments, therapeutic services, and psychometric assessment opportunities, all of which have proliferated throughout the Western world since Positive Psychology first emerged a quarter of a century ago (van Zyl et al., 2024).

In short, Positive Psychology is framed as a science of human happiness, but the narrative it imposes is politically contrived: it instructs people how to be happy in a manner convenient to those who enjoy pre-existing privileges in society. It promotes positive thinking as a remedy not to address poverty per se, but to assist the poor to 'at least be happy in poverty' (Moghaddam, 2023; p. 81). Mindfulness, flow, and optimism? *Yes.* Overthrowing the capitalist system? *No.*

The Positive Psychology approach contributes to a cultural climate that infantilizes, rather than emancipates, its stakeholders (Fernández-Rios & Vilariño, 2016). It highlights how psychology's individuocentrism reinforces ideas consistent with the traditional conservative view of human beings as self-reliant individuals who can each succeed if they really want to, with no need to blame society for human suffering or to interfere with its arrangements in an attempt to address social ills.

30

Example #5: By 'Othering' Ethnic Minorities

Among those being told that happiness comes from within will be victims of discrimination, abuse, and marginalisation. Under the rubrics of Positive Psychology, socially excluded people are urged to seek 'benefit' within their predicaments—literally, to work on *themselves* rather than on tackling society. This advice not only legitimises out-group hostility as a to-be-expected cultural norm, it shifts responsibility for the pain of marginalisation onto the marginalised themselves. If people are distressed at being discriminated against, it is implied, it is because they have failed to successfully utilise self-care remedies that would have provided them with a more 'positive' outlook. A perverse effect of the commodification of happiness is to rebuke the traumatised for having insufficiently optimistic mindsets.

The cultural inwardness of Positive Psychology reflects the broader individuocentrism of psychology as a field—it ignores the political contexts that lead so many people to feel 'negative' about their lives. Moreover, its emphasis on individuals reflects a largely Western outlook. Cross-cultural researchers have long examined how perceptions of human interdependency vary around the globe. In much of Europe and North America, cultural values tend to prioritise personal freedom

B. M. Hughes, *Psychology's Quiet Conservatism*,
https://doi.org/10.1007/978-3-032-07724-0_30
205

over social obligation, and human beings are portrayed as having a natural need for autonomy. However, in many other parts of the world, the dominant view is that personal gratification cannot be divorced from relationships within communities, and that collective needs intrinsically take precedence. In this sense, the individuocentrist stance of psychology reflects a narrow set of Western interests—in other words, it reflects the worldviews of predominantly White cultures—with the direct consequence that it is frequently insensitive to the perspectives of *non*-Western (or non-White) persons.

We will examine this overall cultural skewness—the Whiteness of psychology—in detail in Part VI. However, for the moment, it is worth noting psychology's intrinsic proneness for conservative views of ethnicity. In reality, the field of psychology, time and again, provides fodder for those who wish to advocate for anti-progressive attitudes toward minorities, multiculturalism, and migration.

Individuocentrism as Othering

Psychology's ethos of individuocentrism resonates with the social conservative take on race and racism. Conservatives argue that racism is exclusively a problem of present-day individual prejudices and toxic interpersonal behaviours. They deny the reality of *structural* racism— the demonstrable proposition that many of our societies' institutions, laws, and practices have evolved in ways that persistently disadvantage some people based on 'race' (e.g., Lindsay, 2020). They reject claims that racism has arisen from a history of unjust treatments of minorities by majorities, such as local community segregation, global colonialism, or imperialism. For conservatives, racism is caused by people-who-are-racist and not by historically shaped cultural and political systems. They assert that *remedies* for racism should therefore focus on *altering the behaviours of individuals* and not on modifying, much less overhauling, society. When reflecting on the roots of racism, conservatives prefer to focus on person-level factors rather than on culture, history, or politics.

In short, the conservative analysis of racism adopts, and exploits, the individuocentrism of psychology.

Once again, it is important to note that few psychologists see individuocentrism in political terms. Rather, psychologists tend to believe that focusing on individuals makes their research scientifically more robust. It helps them to extract the 'signal' of personal experience from the 'noise' of life's milieu. However, when it comes to problems such as racism, the result of their approach is to divorce psychology from sociocultural reality. By focusing on person-level phenomena—such as stereotyping, projection, fear, authoritarianism, or meta-cognitive self-reflection—the psychological literature on racism separates this complex social harm from its origins in systematic oppression. It lets bystanders off the hook. It encourages a corporate-friendly version of anti-racism training that keeps attention on trivia such as unconscious bias and empathy, and discourages people from considering how they might go about dismantling institutional racism or resolving political injustice. In the modern world, under psychology's influence, the idea that racism is best dealt with in a workshop setting has been normalised.

Reducing Race

Psychology's proclivity for reductionism creates similar problems. At a basic level, relying on psychometric tests to produce single-number summaries of arbitrarily selected attributes in culturally heterogeneous populations is the very essence of racial reductionism—it takes diverse collectives and repackages them as bland blobs of human indistinctiveness. But more subtly, the process involves the over-generalising of favoured frames of reference to the benefit of dominant groups.

Psychometric assessment instruments often include content that is perceived differently by people from different cultures, in ways that render the instruments culturally insensitive. Differences in scores between cultural subgroups are likely to stem from dimensions of test content rather than from psychological variation across the people being tested. Despite this basic shortcoming, such instruments are frequently kept in regular use. One example is the Boston Naming Test, a neuropsychological tool for assessing aphasia. The BNT requires participants to name the objects in a series of pictures, one of which depicts a noose (Buchanan et al., 2021). Given this item's particular association with racially motivated hate crimes in the United States, it is very likely to have specific resonance for Black people and so to distort the meaning of their scores on the BNT. It is well established that the BNT works just as well when the noose picture is not used (Salo et al., 2022; Zimmerman et al., 2022), and yet the official test has never been updated. Ultimately, it appears that psychologists will persist with an imperfect tool so long as it provides numbers that can be crunched. Measurement 'utility' is allowed to trump cultural insensitivity because, ironically, such insensitivity will only ever affect a *minority* of participants.

Reductionism also leads psychologists to discuss racial identity in overly simplistic ways. In psychology research, the race of study partici-pants is frequently represented by a single data-point, with each partici-pant classified as belonging to one of a small number of categories. Rarely is the list of categories that exhaustive. Commonly, for example, every person with African or Caribbean ancestry is classified, simply, as 'Black'. Another typical category heading is 'Asian', referring to a region that expands across eleven of the world's twenty-four time zones. Efforts to produce separate Asian subcategories are usually rudimentary, despite the fact that Asia comprises several dozen countries and is home to more than two thousand distinct linguistic communities. And it is almost unheard of for questionnaire race categorisations to appropriately accommodate people who have more than one ethnic identity, such as where a person identifies as both 'Black' and 'Asian'. Such inability to recognise diversity of heritage is reminiscent of the way social conserva-tive commentary frequently questions, and thus demeans, the very idea of mixed-race identities (as exemplified by political attacks directed at the 2024 US presidential candidate Kamala Harris, who was mocked by her opponent for 'being of Indian heritage' but willing to 'turn Black' for political purposes; Sullivan, 2024).

The use of reductionist categories to describe people's race and ethnic diversity serves to hide that diversity, and ultimately to render it invisible. In psychology journals, some three quarters of research papers fail to provide any data at all to describe the race or ethnicity of study participants (DeJesus et al., 2019). For research conducted in White-majority countries, psychology's reductionism of race and ethnic identity is especially problematic, because it reflects the minimising way in which minority populations are depicted by those in society who are most hostile to them. Summarising minorities using simplistic categorisations involves a style of out-group homogenisation that resembles that frequently seen in xenophobic rhetoric. It suggests that all readers need to know about these people is that they are not White.

Conventions of Exclusion

As we will explore in Part VI, mainstream psychology effectively presumes the standard human being to be White. This is the case even though that label can reasonably be applied to just seven per cent of humankind. Further, the vast bulk of psychology journals are published in White-majority countries (chiefly the United States), with the result that most research participants being reported on are White people. Not only that, but the reporting conventions of these journals are manifestly White-centred.

Studies focusing on non-White participants are less likely to be published unless a White comparison group is included in the protocol and the race of participants is indicated in the title of the article (Wang, 2016). Meanwhile, studies that focus on White people are almost never required to incorporate comparisons with non-White controls or to mention participant race in their titles (Roberts et al., 2020). Standard practice in psychology communicates that Whiteness is the human default, and that non-White people are to be discussed in terms of how they deviate from White norms, essentially framing non-Whiteness as an aberration. Even though the reporting standards of academic psychology journals strive for scientific neutrality, they result in practices that sometimes, to be frank, are difficult to distinguish from the discourse of White supremacism.

Of course, when considering such conventions, we must again recall that psychology's main psychometric methods and statistical tests were devised for the purposes of eugenics. Even the basic approach of classifying human beings into comparison groups, a practice beloved by psychology researchers, can be traced to this dubious heritage. Funnelling participants into two or three groups—each represented ultimately by a single summary statistic—is the very epitome of reductionism. Human beings are, if anything, characterised by their diversity; even identical twins will develop diverging interests and personalities. Grouping participants necessarily elides diversity *within* groups and exaggerates differences *between* them—even when those differences are fluid, trivial, statistically non-significant, or falsely significant as a result of some methodological artefact or statistical fluke (that is, even when they constitute 'false positive' findings). With its illusory accentuation of so-called group differences, psychology's reductionism provides an ample stream of material for those whose political objectives demand scientific-sounding talking points with which to demonise or scapegoat marginalised communities.

Of course, not all group-based reductionism is hateful. For example, some race stereotypes appear at first glance to be superficially flattering, such as the idea that particular ethnic groups can be regarded as 'model minorities' because of their positive contributions to society. But while not conspicuously toxic, these tropes are nonetheless harmful (Yip et al., 2021). They are patronising, one-dimensional, and ultimately based on anecdote rather than evidence. And they implicitly judge the value of the out-group with reference to the needs of the in-group, such as whether the minority successfully participates in the majority's 'society'. They thereby maintain the implication of in-group standard (and dominance) in contrast to out-group deviance (and subjugation).

Individuocentrism, by definition, ignores political context and social responsibility. And for its part, reductionism ultimately necessitates reduction. It involves the loss of detail, the removal of knowledge, and the degradation of understanding. It serves no interests other than those of people who *want* such knowledge and understanding to be reduced, lest we find ourselves developing empathy with those we are told are fundamentally different to us. The methodological conventions upon which psychology relies ensure that the field's knowledge-narratives consistently 'other' ethnic and race minorities. These narratives minimise the role of structural and cultural racism, render invisible the nuances of ethnicity and race, and imply that White people represent the 'pure' form of human psychological existence. They echo, and reinforce, socially conservative views on ethnic diversity. Such an outcome is hardly the attribute of a field that can be described as 'liberally biased'.

31

Example #6: By Policing Gender Identity

As we saw in Part I, many culture war commentators have amassed significant public profiles by pontificating on trans issues. Questions of gender identity regularly spark campus conflicts, with positions sympathetic toward transgender people cast as insufferably woke, and anti-trans scepticism becoming something of a conservative cause célèbre. While hating on trans people has a long history within social conservatism, it has latterly become a burning touchpoint in cultural politics. Conservatives typically frame transgender identity as pathological, decadent, and inconsistent with what they unilaterally deem to be 'biological' (or 'scientific') reality. In today's culture wars, being sufficiently insulting toward trans people as to be cancelled can yield lucrative conservative martyrdom, if not right-wing immortality.

By contrast, most psychologists tend to be personally supportive of transgender people, and it is typical for professional psychology organisations to vocally champion trans people's rights and dignity. Often these stances are promoted in direct response to conservative anti-trans policies that restrict the freedom of trans people to live according to their experienced genders. In 2024, the American Psychological Association issued a major position statement affirming the human rights of trans

B. M. Hughes, *Psychology's Quiet Conservatism*,
https://doi.org/10.1007/978-3-032-07724-0_31

individuals. The declaration explicitly condemned political attempts to scapegoat people who are transgender, gender-diverse, or nonbinary (APA, 2024). That same year, the British Psychological Society published new guidelines on gender, sexuality, and relationship diversity in adults (BPS, 2024b), advising its members that 'diverse gender identities are a normal part of human diversity' (p. 5), that 'diversity of genders . . . is something that psychologists acknowledge and support' (p. 7), and that 'attempting to change or suppress a client's gender or sexuality as a primary goal of treatment is unethical' (p. 6). But while many professional groups have championed trans-inclusive values at a corporate level, the positions taken by representative organisations are not always universally endorsed. Trans allyship can sometimes be stridently opposed by individual member psychologists (e.g., in the UK, Pilgrim, 2024).

But once again, when it comes to gauging the politics of psychology as a field, the positions taken by individual psychologists are seldom the key indicator. As we have seen, psychologists often see themselves as thoroughly progressive while remaining oblivious to their field's promotion, and preservation, of socially conservative ideas. Statements by organisations in support of trans rights might signal virtue to liberal audiences, but they also distract attention from the way psychology's conventional customs continue to reinforce negative and marginalising stereotypes about gender-diverse people.

Disease Models

One problem relates to how psychologists frame their theories about human bodies. We saw in Part III that psychologists tend to gravitate toward biopsychosocial theories in which the physical is explained by the psychological—for example, theories that blame ulcers on stress, migraine on perfectionism, or long COVID on mass hysteria. Conservative arguments about transgender people evince a similar impulse to psychologise. From the conservative point of view, transgender status should not be seen as an authentic biological reality, but as something self-reported and thus rooted in psychological processes. Once again, physiological function is overlooked in favour of the superordinate psyche. Assuming binary gender to be the 'scientific' default, they conclude that transgender *cannot* therefore exist. As such, any claim to deviate must be 'all in the mind.' Transgender identities should be assiduously ignored or, if that is not possible, demeaned as mental illness.

Psychology's theories about human bodies complement these socially conservative tropes. Anyone committed to binary-gender intransigence can take comfort from paradigms that attach pre-eminence to the human psyche. They can draw on psychology for inspiration when dismissing transgender identities as mere in-the-moment feelings, as cognitive templates socially imposed on physical reality, or as conceptual errors perpetrated by people whose minds are malfunctioning.

Psychology's individuocentrism exacerbates all these problems. It frames human predicaments as if they are determined entirely by a person's own actions and choices. Crucially, it ignores how a person's actions and choices are influenced by those of *other people*. This is especially pertinent for anyone who is trans, given that their very identity as 'trans' results from the fact that *other people* assigned them a particular sex at birth with which they do not now identify. Further, many of the difficulties faced by trans people stem from the ongoing obsession that *other people* seem to have with their gender. It is *other people* who question their presence in bathrooms, libraries, and sports arenas. It is *other people* who protest when they are provided with gender-affirming medical care. It is *other people* who object to their existence being acknowledged in sex education curricula. And it is *other people* who make a point of refusing to use their preferred pronouns. Indeed, the entire idea of transgender is often meaningful only in this context: were it not for the persistent reactions of *other people*, we would have few reasons ever to refer to transgender as a concept.

For many decades, the field of psychology was embedded among those 'other people' drawing negative attention to transgender lives. The conventional view among psychologists was that trans people suffered from a psychiatric pathology, namely, the aforementioned gender identity disorder or GID (Winters, 2011). Such language of 'disorder' clearly has the potential to offend, and on its own it was responsible for the emergence of much of today's transphobic discourse.

In 2013, GID was removed from DSM-5 and replaced with 'gender dysphoria', the criteria for which include distress associated with an incongruence between one's gender identity and one's birth-assigned sex, as well as 'a strong desire' both to *be* another gender and to be treated as such. All of these could be said to apply to nearly every person who is trans. What the criteria do *not* seem to account for is the fact that when distress arises, it is often not because of a trans person's own being, but because of how they are treated by *other people* (including legislators and bureaucrats). Classifying a person as 'gender dysphoric' based on their own feelings is squarely individuocentric. It fails to acknowledge the possibility that transgender people would be just fine were they not continually impeded, harassed, and abused by an inherently transphobic society. As noted earlier, the entire enterprise of psychiatric diagnosis is open to the same criticism. Individuocentrism drives narratives that blame victims more than bystanders.

Psychologists often proclaim the DSM's shift from *gender identity disorder* to *gender dysphoria* as a progressive evolution, given its removal of the term 'disorder' from the discussion. However, it is debatable whether the new terminology significantly lessens the stigma. Firstly, the word 'dysphoria' is still premised on the presence of a (mental) *dys*function; indeed, the term merely specifies the *type* of dysfunction these people are supposed to have (namely, one relating to emotion). It is just as problematising as any reference to a 'disorder'. Secondly, DSM-5 continues to associate gender dysphoria with significant distress in a person's life, thereby implying an inherent overlap between gender diversity and poor mental health. And thirdly, the very fact that gender dysphoria appears in the DSM at all presents an issue. After all, the DSM is a publication of the American *Psychiatric* Association. Its full title is the 'Diagnostic and Statistical Manual of *Mental Disorders*'. So, while gender dysphoria is said not to be a psychiatric disorder, it is nonetheless listed in the official directory of disorders used by professional psychiatrists.

The DSM's international equivalent—the World Health Organization's International Classification of Diseases (ICD)—provides a slightly different approach, referring to *gender incongruence* instead of *gender dysphoria*. In addition, the ICD does not specify mental distress as a defining feature for gender incongruence, and it no longer includes gender incongruence in its 'Mental and behavioural disorders' chapter, but instead in its chapter on 'Conditions related to sexual health'. However, despite all these precautions a glaring fact remains. The clue is in the title. The ICD is a system for classifying *diseases*. If anything, the ICD stance could be said to be even more problematic than that of the DSM. From the ICD perspective gender incongruence alone is worth classifying even when no distress is present. In other words, transgender people who are *in perfect mental health* are nonetheless said to have a status worthy of inclusion in a directory of 'diseases'.

Oppressive Norms

Social conservatives often employ disease terminology when referring to trans identities. For example, in 2025, when attempting to ban trans people from the US military, the conservative American administration framed its key anti-woke policy as covering service personnel who 'exhibit symptoms' of being transgender (Reed, 2025). Similarly, the Cass Report, a highly controversial but nonetheless influential review of gender identity services in England, makes extensive use of pathologising language when describing trans adolescents and their lives. Notably, it makes no reference at all to the existence of 'trans children', but instead frames youth transgender as a disease, at one point likening gender diversity to cancer (McNamara et al., 2024). One consequence of employing such a medicalised frame of reference is to create a heightened sense of threat to public welfare, and thus to distract from significant weaknesses in the research that the Cass Report presents to justify its anti-trans conclusions (e.g., Noone et al., 2025).

But psychology's own discourse on trans issues also accommodates much pathologising language. Even in trans-affirmative contexts, reference is frequently made to the 'prevalence' or 'aetiology' of transgender identity (e.g., Spivey & Edwards-Leeper, 2019), as though what is being discussed is an epidemic infection that requires a public health response. And a cursory literature search on how transgender topics are presented in psychology journals will reveal a focus on trans people's mental health challenges and adverse life experiences, such as low self-esteem, minority stress, depression, and suicide (e.g., Pellicane & Ciesla, 2022), as if transgender and misery automatically go hand in hand.

Even more revealing is that part of the psychology research literature that does *not* deal directly with trans issues. After all, psychologists who write about trans people will—rightly—have their language closely scrutinised and, in the context of peer-review, improved where necessary in order to ensure its inclusiveness. However, when research is ostensibly unrelated to transgender people, or to gender issues more broadly, that vigilance is seldom applied. Such research potentially provides a more effective lens through which to observe the field's predisposing biases.

Similarly to what we saw with race and ethnicity, psychology research tends to take a rather simplistic—i.e., reductionist—approach when classifying participants by gender. Gender is almost always presented as if it were a binary concept. Dropdown menus on study questionnaires frequently offer only two gender options, requiring trans, nonbinary, and gender-fluid individuals to subsume themselves into categories in a way that might not have been intended (Garb, 2021). A likely consequence is that many trans individuals will be reluctant to participate in psychology research altogether, either because they feel uncomfortable answering the gender questions they are presented with, or because they wish to avoid having to 'out' themselves to the researcher (Kiperman et al., 2021). Further, many standard measures are ill-suited to participants who are transgender. Several psychometric tests employ different scoring scales for females and for males, with transgender people not identified as distinct and without it ever being clear whether the intended contrast can reasonably be generalised from cisgender to transgender participants (Webb et al., 2016). And even when gender diversity is recognised by researchers, it is often reduced to a single data point, with widespread conflation of transgender identities into that of a supposedly homogenous

'LGBTIQ + community' (Kiperman et al., 2025). In reality, the majority of samples described using this heading consist mostly, if not exclusively, of cisgender individuals (APA, 2021).

The overall impact of standard methodological practice is to ensure that in most psychology research, trans people are likely to be misrepresented, underrepresented, or rendered invisible.

Trans Panics

Reductionism in psychology also facilitates the formation of moral panics around gender identity. As first famously described by the sociologist Stanley Cohen (1972), classic moral panics involve the identification of scapegoats (or 'folk devils') and the exaggeration, or fabrication, of controversies. This provides rhetorical space for society's power brokers to implement disproportionately negative responses toward marginalised or easily victimised groups (who, by definition, are largely unable to defend themselves), as a way of demonstrating their moral authority and capacity for decisive action (Schneider & Ingram, 2005). Moral panics involve the harnessing of emotional energy in order to influence the social dynamic (Young, 2009).

Key to the process is the ability to reduce complex issues into a set of simple symbols. It is very hard to whip up anger over the nuances of genetic biology, or the multifactorial processes whereby young people come to perceive and understand their gender and sexuality. It is far easier to stir up anger using simple examples that the public can readily visualise. In modern moral panics, reductionism is all the more important given the centrality of social media sound bites within political communication (Pepin-Neff & Cohen, 2021), which naturally thrive when topics are suitably simplified. It is for such reasons that modern conservatives exhibit a near-obsessional interest in gender-neutral bathrooms, trans athletes, Drag Queen Story Hour, trans prisoners in jails, trans-Pride flags on public buildings, and, without wishing to be crude, (other) people's genitalia. This is where psychology's role-modelling of reductionism becomes useful.

Psychology's approach of eliding complexity in ways that render trans people barely visible, and therefore inherently odd, helps to bolster such moral-panic reasoning. Its lingering intonation of trans-specific psychopathology further amplifies cisnormative anxiety around the existence of transgender lives. And its customary focus on individual perspective draws attention away from the social, cultural, political, and historical aspects of gender—and indeed of transphobia itself—thereby reducing 'debates' on trans matters to the lowest of common denominators. Notwithstanding the undeniable trans allyship expressed by many individual psychologists (and by their representative bodies), psychology's reliance on individuocentrism and reductionism as methods of knowledge production directly contributes to negative discourse on trans issues, in no small part reflecting—but also reinforcing—cisnormative prejudice in wider society.

32

Example #7: By Perpetuating Traditional Gender Stereotypes

In addition to viewing humanity as comprising just two fundamental types—females and males—socially conservative ideologies also subscribe to gender essentialism, the idea that women and men have inherent, biologically determined traits that shape their psychologies. Traditionalist conceptions of gender have long held that women and men differ sharply in their temperaments, tastes, behaviours, sociability, and aptitudes. Moreover, the belief is that for the world to be right, women and men *must* personify these differences. Women, for example, must be thin, fertile, and make sandwiches (Badham, 2025). This is why anti-trans sentiment is so deeply socially conservative. Much of the condemnation of trans people is rooted in the fact that trans women tend to be insufficiently 'feminine' to satisfy conservative sensibilities. Their very existence challenges conservative stereotypes about what women are supposed to be and how they are supposed to behave.

Whether believed to be endowed by deities or predetermined by evolution, these supposedly fundamental sex differences are cited by conservatives to justify the separation of men and women's social roles. Men, it is argued, are naturally tuned for competition, construction, and commerce. Women, on the other hand, are said to be primed for

B. M. Hughes, *Psychology's Quiet Conservatism*,
https://doi.org/10.1007/978-3-032-07724-0_32

nurturing and domesticity. The depiction of women and men as psychologically distinct is not sexism, conservatives argue, but *science*.

The field of psychology has long shared the conservative fascination with human sex differences. But it is far from clear whether its scholarship has uncovered an authentic divergence in the psychology of women and men, or whether it has simply drawn from, and bolstered, false sexist stereotypes. There has certainly been a concern that psychologists are liable to go about this type of research with especially poor rigour. 'There is perhaps no field aspiring to be scientific,' wrote the psychologist Helen Thompson Woolley when reviewing the psychology of sex differences in the pages of *Psychological Bulletin*, 'where flagrant personal bias, logic martyred in the cause of supporting a prejudice, unfounded assertions, and even sentimental rot and drivel, have run riot to such an extent as here.' That Woolley offered this damning verdict more than a hundred years ago (Woolley, 1910) is especially striking. Her withering takedown of this persistent genre of pseudoscience could just as easily have been written today.

Martyred Logic

Unfounded prejudices and sentimental rot and drivel have continued to shape psychology's theorising about sex differences. We have already discussed a number of examples, such as the view that women are more likely than men to succumb to psychogenic illness. It is worth noting how insidiously such stereotypes cohere with sociopolitical demands. Not all women are said to develop these hysterical maladies, only those whose personalities are potentially problematic for men. Women who exhibit ambition, perfectionism, or manipulative tendencies—those who are likely to intrude into male-dominated domains of industry and power—are the ones said to be at risk. Women who *lack* ambition should be fine.

Fear that female empowerment would, in fact, be dangerous for women (and thus for all of society) has been a long-running sexist trope. In the eighteenth century, Christian philosopher Mary Astell wrote of how high-achieving women were widely perceived as having 'acted *above* their sex' (Astell, 1705; italics added). In the nineteenth century, a board member at Harvard Medical School warned that women who entered college risked being rendered infertile due to the significant blood flow that academic thinking would require of feminine brains (Clark, 1873). In the twentieth century, pioneering psychologist G. Stanley Hall pointed out that the 'natural differentiation' between the sexes made female ambition especially inadvisable, noting how women's eagerness 'to abandon home for the office or shop, or to strive for intellectual careers ... is hard upon their constitution' (Hall, 1906). And well into the twenty-first century, psychologists are still wont to examine whether excessive perfectionism might in fact be problematic for women, by exacerbating their symptoms when they are chronically ill (e.g., Deary & Chalder, 2010).

Modern presumptions about the social benefits of gender roles—that is, about the (proper) place of women—are sometimes so subtle as to slip by unnoticed. However, in the past, claimed links between sex differences and societal well-being were often voiced explicitly. Nineteenth-century anthropologists argued that patriarchal societies represented the pinnacle of civilisation, and that prehistoric tribes in which women were in charge had become extinct due to their inherent shortcomings (e.g., Bachofen, 1861). Psychologists in the early twentieth century shared the view that sex differences reflected, and indeed drove, the march of modernity. 'In savagery, men and women are more alike,' Hall wrote in 1906, '...but with real progress the sexes diverge' (p. 590). The overall argument was that traditional gender roles reflected the optimal arrangement for the human species, with the differing psychologies of women and men operating symbiotically to enhance the viability of civilisation. An important consequence of this reasoning was the warning that changing this arrangement risked potential catastrophe. Denying the necessity of psychological sex differences—or, even worse, attempting to *undo* them—would, by flying in the face of nature, threaten humanity itself.

Reproductive anxiety transcended academic psychology in its formative years. Hall decried a nightmare scenario populated by spinster women whose 'over-activity of the brain' would 'interfere with the full development of mammary power' and bring to an end 'the functions essential for the full transmission of life generally' (1906; p. 592). To this day it continues to permeate broader political commentary, with twenty-first century conservatives frequently echoing their own concerns about the perils of female childlessness. In 2016, a candidate for leadership of the UK conservative party said of her (female) opponent that she must be 'really sad she doesn't have children', citing this unnatural status as a sign the woman lacked a 'very real stake in the future of our country' (Coates & Sylvester, 2016). A more recent—and blunt—example was the American vice-presidential candidate who described progressive woman politicians as 'a bunch of childless cat ladies who are miserable at their own lives', provoking a storm of media criticism. He went on to complain that it was dangerous for his country's future to be 'controlled by people without children' (Pengelly, 2024b). Despite widespread condemnation of these remarks, his point appeared to resonate with the wider public, who in due course rewarded him with electoral victory in the 2024 general election. After a century or more of repetition, claims about the biological reality of sex roles have become embedded in popular conservative beliefs about women and men.

The mainstream psychology research literature on supposed sex differences tends to conform to the conservative worldview. Men are portrayed as having particular aptitudes for numeracy, leadership, systems analysis, and visuospatial reasoning, and as having more self-confidence than women. By contrast, women are described as being less aggressive than men, as well as more nurturing, less likely to cheat, more talkative, and less impulsive (Hyde, 2014). Even the classic human fight-or-flight response has been hypothesised to differ by gender in a stereotyped way: women are said to have their very own stress response that eschews men's aggression and evasiveness, instead reacting to moments of crisis by instinctively engaging in *tending-and-befriending* (Cohen & Lansing, 2021). The theory suggests that when times are hard, men and women's natural skills come to the fore: men are biologically primed to react directly to problems, whereas women are naturally driven toward caregiving roles.

The implicit message within all these ideas is that, for good or for ill, men are just better suited to managing the things that impact our lives. Women, on the other hand, are better suited to supporting roles, such as providing empathy, tending to social ties, or looking after those who require care. The social and economic consequences of choosing to defy these supposedly immutable sex differences are rarely highlighted in psychology journals. But the inherent political argument is easy to extract.

Many psychologists see themselves as feminists. However, this does little to change the fact that their field provides a torrent of material for conservative case-making, such as when men's natural leadership abilities are contrasted with women's precocity for nurturing and empathy. The problem is that the empirical evidence said to underlie these purported truths is far from robust. As shown repeatedly by University of Wisconsin psychologist Janet Shibley Hyde (2005; 2014; 2018), any systematic assessment of the data will reveal claimed sex differences in psychological attributes to be hugely exaggerated. Most statistical differences are either non-existent or trivial. Those that are not usually cannot be replicated, or are demonstrable only in certain cultures. And the very fact that sex differences can be shown to vary by culture pretty much proves they are *rooted* in culture rather than in sex. As Hyde points out, there is much better evidence in psychology for a generalised 'sex similarities' hypothesis than there is to justify the field's current obsession with hunting for (or gathering) sex differences (Hyde, 2014). And yet psychology's fascination with sex differences seems unremitting. A century on, Woolley's description of a literature laden with flagrant bias, prejudiced logic, and sentimental nonsense would appear to remain valid.

Standard Deviations

A particularly pertinent example for psychologists to consider is the purported sex difference in mathematical ability. According to the common stereotype, this is related to inherent biological factors that make men more likely to succeed in numeracy-based careers. In its simplest form, the view has it that women simply find mathematics more difficult, as famously immortalised in the now discontinued Barbie doll toy that incorporated a voice box declaring to young girls that 'Math class is tough!' (Associated Press, 1992). However, despite what Barbie might have to say, extensive empirical evidence suggests that this stereotype is untrue. Data from literally millions of school students show the overlap between girls' and boys' arithmetic performance to be around 99.9% (Hyde et al., 2008), with the remaining variance almost certainly attributable to statistical 'noise'. The random nature of this noise is revealed by the fact that the *direction* of statistical difference varies across the data: in some age-groups boys slightly outperform girls, but in others, girls slightly outperform boys. A true sex difference would not switch haphazardly by cohort. Similar studies of adults produce precisely the same conclusion (Lindberg et al., 2010).

Those psychologists still committed to arguing for a sex difference in mathematics can instead turn to a more subtle form of the stereotype, which we encountered in Part I. This is the variability hypothesis, the claim that men's mathematical performances *vary* more than women's, a pattern that would naturally create a preponderance of men at *both* extremes of the ability scale. As we described previously, it was this idea the former president of Harvard had in mind when he claimed that it was entirely understandable for elite universities to end up hiring more male scientists than female ones. However, the variability hypothesis also becomes untenable once the evidence is consulted. The relevant statistics suggest that even at its most extreme, the imbalance of men and women in the top five per cent of mathematics scores is likely to be no worse than 52% versus 48% (Hyde, 2014). Such a tiny difference in variability simply cannot explain why only 18% of undergraduate engineering degrees are awarded to women.

What makes this pertinent for psychologists is the fact that professional psychology itself appears to have bought into this stereotype. Often the point is made in terms of interests rather than abilities, in that female and male psychologists are regularly presumed to have different levels of attachment to specific types of research. For example, the British Psychological Society's Standing Committee for the

Promotion of Equal Opportunities once argued that sex differences in quantitative reasoning helped explain why male psychologists were more likely to be favoured by professional psychology's incentive systems. Quantitative research methods, they said, were likely 'to reduce women's participation in psychology', presumably because women were less likely than men to want to 'participate' in this type of research (Riley et al., 2006; p. 96). The reason *why* women were less inclined toward quantitative research was not stated. But as people usually prefer to engage in activities in which they are proficient rather than ones in which they flounder, the allusion to a sex difference in quantitative *ability* was difficult to ignore. As if to reinforce the point, the Committee went on to praise the BPS's newly established qualitative methods subgroup as 'an important forum' for addressing institutional sexism, on the basis that it would provide recognition 'for those who adopt' qualitative techniques (in other words, women). The message was clear: women and men's different experiences in the psychology workplace were related to the demands of quantitative methods. The pre-eminence afforded to statistical reasoning in psychological science was unfair. Quantitative methods create a *gendered* glass ceiling. Math class is tough.

But not only is this stance iffy on empirical grounds, it is also damaging for reinforcing socially conservative views on problems such as structural bias, by focusing on person-level factors rather than on social, cultural, historical, political, or economic forces. In reality, the barriers encountered by women psychologists in academic settings emerge from a wide range of interconnected injustices, including stereotypes *promoted by other people* over which they themselves have limited control. Blaming quantitative methods is a distraction, and a stigmatising one at that. It portrays women psychologists as deficient—as lacking the core skills necessary for academic success.

This is yet another example of how individuocentrism shifts blame onto victims by focusing on (purported) personal deficits rather than contextual drivers of inequality. Except, in this example, the victims are not the participants of psychology research, but psychologists themselves. An irony here is that qualitative methods continue to be frequently defended by feminist psychologists precisely because of their perceived gender 'relevance', a tactic that seems to try to emancipate women by casting them (at least implicitly) as less numerate than men. A second irony is that this entire cultural stereotype has remained rigidly persistent despite the availability of reams of contradictory *statistical* evidence. In a patriarchal world, men often get to decide what the problems really are and what really causes them; if men really *do* believe in mathematical sex differences, perhaps they should brush up on their *own* numeracy and learn how to interpret the relevant statistical evidence more effectively.

As it happens, psychology's continuing tolerance for sex-difference stereotypes is itself directly traceable to deficient quantitative reasoning. Much of psychology's widely discussed replication crisis, which we will discuss in Part VII, could be said to stem from the fact that psychologists, on the whole, have a poor grasp of the statistical principles upon which their field's research is built. In short, there appears to be widespread confusion over the meaning and computation of probability. A consequence has been an emerging trope that sees p-values treated as a proxy for scientific replication. Unlike other scientists, psychologists rarely ever conduct replication studies, partly because they place too much store in p-values: if something was significant before, they feel, then why should we test it again? But in reality, as we will see later, the true rate of spuriously significant findings in psychology is gravely underappreciated. Indeed probably *more than half* of published findings are liable to be false positives, random outcomes that appear statistically significant but which in reality are not (Ioannidis, 2005).

Coupled with this is psychology's odd custom of insisting on reporting the (binary) genders of all research participants, even if the given research question has nothing to do with gender or sex. Sex differences in all sorts of minor and core variables are unthinkingly analysed in blanket checks that psychologists consider their inherent duty to report. This makes sex differences the most tested category of hypothesis in all of psychology, despite the fact that the vast majority of research studies, at least on their face, are designed to investigate other issues. As some 200,000 papers appear in psychology journals each year—equating to 500 new papers per day (or one every three minutes)—it is little wonder the field has accumulated a huge repository of opportunistically computed but spuriously positive statistical test results implying that men are from one planet and women are from another, junk findings that reinforce a host of ancient cultural prejudices while spawning an endless stream of new ones.

Natural Misogyny

Of course many human parameters do differ by sex, such as average height, average weight, proneness to heart disease, membership of elected governments, likelihood of riding a motorcycle, participation in professional rugby, and, in Western societies, average length of hair. However, by and large, in popular culture these differences are typically interpreted in highly simplistic ways, encouraged in part by psychology's proneness to biological reductionism.

For example, just about every criminology dataset shows that men are more likely than women to engage in serious violent crime. Male violence is typically attributed to several physical factors, such as body size and muscularity, as well as to hormones, like testosterone, that supposedly precipitate anger and rage (e.g., Bartle, 2019). The science of psychology is then invoked to explain how these attributes are shaped by evolution through the dynamics of natural selection. The effect of this type of account is to portray the problem of male violence as a product of biology rather than of other factors. Even worse, the biological imperative largely serves to *excuse* male violence as a force of nature, as if men could never be fully in control of their own aggressive acts. Viewing male violence as a component of human biology presents few avenues for remedial intervention. While it is possible in theory to target testosterone using medication (or castration), in practice, biological reductionism offers only incapacitation and punishment as responses to masculine aggression.

Biological reductionism and individuocentrism in psychology encourage people to think of male violence as a 'natural' physiological event that originates within the bodies and brains of individual male perpetrators. As such, they prevent people from thinking about social, cultural, historical, economic, or political factors that either directly or indirectly prompt men to be violent, such as social permissions (for example, male privilege), material incentives (for example, financial enrichment as might be necessitated by economic desperation), cultural radicalisation (for example, incel or rape-joke culture), inconsistent or non-existent punishment regimes (for example, the poor prosecution rates for rape), and gendered forms of cultural socialisation (where men are rewarded with social capital when they are physically aggressive, or ridiculed as a 'lesser' man if they refuse to be violent). Such social norms are themselves shaped by collective attitudes and behaviours, and transmit their effects both directly and vicariously. Men frequently witness powerful figures receiving acclaim for their aggression, such as politicians who attain electoral success after threatening, or even committing, violence against other people. Men who become violent under these conditions need not have high testosterone levels, nor do they have to be physically strong. They just need to *choose* to engage in a violent act, in the knowledge that their society both enables and facilitates male violence, and even rewards it.

The result is that biological factors are typically deployed as an excuse more than as an explanation. A man accused of violence can claim diminished culpability because of hormones outside his control and evolutionary forces that long ago programmed his genes for aggression. And not only is male violence (at least partially) excused as an immutable characteristic of masculinity, there is even an implication that, simply by virtue of being 'natural', such sex-typed behaviours must on some level be healthy, at least for the species. Women too are lumbered with biologically reductive stereotypes that depict them as unable to remain composed due to fluctuating oestrogen, or as naturally primed for child-rearing due to innately high levels of oxytocin.

None of these ideas capture the true contextual causes of men and women's different experiences and exploits. Complex social problems are discussed in terms of personal parameters. It is easier to think about punishing a dangerous man (or to blame his female victim for triggering his sexual impulses) than it is to reconfigure the incentive structures embedded within society that reward men who choose to behave dangerously.

Social conservatism tends to embrace this biologically reductive, individuocentric template whenever it discusses traditional gender roles. Evolutionary psychology is used as a justificatory framework within which to provide a scientific veneer for misogyny, enabling abusers to invoke a preordained biological right of men to exert dominance over women (as communicated by the deliberately provocative slogan, 'Your body, my choice'; Badham, 2024)—all on the pretext of some evolutionary expediency alleged to have affected primitive humans living in caves during the Palaeolithic Period. Despite many scientific criticisms (e.g., Jonason & Schmitt, 2016; Pietraszewski & Wertz, 2022), popular evolutionary psychology continues to enjoy mainstream status within academic psychology. Its scholarship is frequently cited by so-called 'men's rights activists' when espousing their views on incel message boards, gleefully claiming that psychological science has validated their beliefs about women's natural subservience to men (Bachaud & Johns, 2023).

Default Sexuality

A final aspect of traditional gender roles involves the belief that men and women are liable to be sexually attracted to each other. In other words, the traditional presumption is that human beings are straight and not gay. Many observers have noted how the field of psychology exhibits such a long-standing heteronormative bias (Thorne et al., 2019).

Psychology's primary theories concerning romantic attraction and sexual development—as well as those relating to broader forms of social interaction—are uniformly premised on the idea that human beings, by default, seek sexual intimacy from long-term, monogamous, marital, and child-bearing opposite-sex pairings. Gay people's relationships are subject to reductionist framing, viewed primarily in terms of their deviation from the standard of straightness. Romantic love is conflated with sexual desire, because that is what straight people presume to be usual. The 'development' of sexuality is discussed only with regard to that of LGBT+ people; how straight people's sexuality might be said to 'develop' (or be 'caused') is seldom explored. Reductionism also leads psychology to overlook the many demonstrable benefits of same-sex intimacy (such as higher relationship satisfaction than seen in opposite-sex couples, more equitable relationships, and more effective resolution of relationship difficulties; Peplau & Fingerhut, 2007), obliquely patho-logising non-straightness. Individuocentrism too dominates the narra-tive, with a focus typically placed on the mental health of gay people rather than on the forms of homophobia that oppress them.

In a related vein, even the most inclusive parts of psychology often depict sex, whether gay or straight, as an innate need. This presumption of sexuality is itself a presumed gender role that excludes, for example, people with lower libidos or who are asexual. The negative stereotyping of sexual apathy finds its place in clinical practice with the DSM diagnosis of *hypoactive sexual desire disorder*, used when a client exhibits low sexual desire in a manner that causes them to feel distress. However, the exact source of this distress is not always clear. Individuocentrism predisposes psychology to presume that such distress must arise within the sufferer, whereas the person may well simply be upset because they feel they are seen as 'frigid' by other people, including those closest to them. The social expectation that all human beings must conform to a traditional ideal of sexuality is itself oppressive. The so-called 'normal' range of sexual desire is, in fact, culturally designated. When a person experiences less interest in sex than they are 'supposed' to, any distress that follows will substantially depend on how other people treat them. Once again, the conventional focus of mainstream psychology tends towards a more traditionalist value system.

The overarching ideology of psychology presumes that women and men fulfil specific roles that stem from their biological evolution. It also presumes that heterosexuality is the default sexuality of human beings, with the implication that other ways of living bring the usual (personal) risks that arise whenever people deviate from conventional behaviour. In these respects, psychology provides a framework of thinking that reso-nates strongly with social conservatism, a worldview that sees men and women each having their natural place within society, that presumes healthy human beings to be straight, and that allows for the existence of diversity only insofar as it is understood to represent a departure from what is 'normal'. Gender roles are described in psychology in terms of unchanging person-factors that are rooted in biology, impervious to societal and cultural influence.

Notwithstanding the thousands of committed feminists and LGBT+ allies who work as psychologists, and who point to their profession's many contributions to the promotion of gender equality and rights, the field's theoretical paradigms lag far behind the political pronouncements of its representative bodies (Thorne et al., 2019). Its methodological conventions of individuocentrism and reductionism bias psychology toward depictions of human gender and sexuality that are laden with conservative assumptions.

33

Example #8: By Exceptionalising Humanity

Psychologists have long seen themselves as scientists of the human mind. William James, the founder of US academic psychology, famously launched the new discipline as 'the Science of Mental Life' (1890; p. 1). As time went by, and ideas evolved, the field's humanitarian ambitions heightened. Pioneering neuroscientist Robert Ornstein saw psychology as so all-encompassing as to constitute 'the study of human experience' (Ornstein, 1988). Today, the website of the British Psychological Society presents psychology as a science that studies 'the mind and behaviour' (BPS, 2025), while that of the American Psychological Association frames it as 'an autonomous scientific discipline' that produces knowledge 'about the nature and development of human thoughts, emotions, and behaviors at both individual and societal levels' (APA, 2020). Uniting all these lofty descriptions is an empirical-explanatory endeavour focused on the profundities of our species: a mission to leverage the methods of science to better understand how humans think, behave, and thrive. In other words, the point of psychology is to provide evidence-based accounts of what being human actually *feels* like.

B. M. Hughes, *Psychology's Quiet Conservatism,*
https://doi.org/10.1007/978-3-032-07724-0_33

Many definitions of psychology acknowledge the study of non-human animals as well as human ones, and there is a long history of psychologists engaging in ethology and other types of animal research. However, aside from some niche exceptions, the majority of these studies are aimed at establishing fundamental principles that can be generalised *from* animals *to* human beings. Such research frequently focuses on isolating the rudiments of common psychological processes that are shared across species, such as learning, motivation, emotion, memory, and social dynamics. Many psychologists choose to study animals in the belief that animals have comparatively less complex brains and nervous systems and so provide simpler models of how biology affects the behaviour of living organisms. Others study animals because they want to employ methods that would be prohibited for use with human participants under current ethical standards. While some psychologists spend their entire careers studying non-human species, their findings are held to be of value—by wider society, by government funding agencies, and by psychology students in universities—only insofar as they inform our understanding of *human* affairs.

So when psychology is described as 'the science of mental life', it is taken to mean the science of *the mental life of humans*. When it is framed as 'the study of mind and behaviour', it is the mind and behaviour of *human beings* that is being referred to. It is *humans* who have personalities, cognitions, mental illnesses, languages, relationships, and so on, in the ordinary senses of these terms. Within psychology, human mental life is placed on a pedestal, reflecting the pre-eminent privilege attached to humanity within most people's very conception of nature.

The elevation of humanity is built into the conventions of psychology. As we saw in Part II, modern psychology emerged from roots in (Western) theology, class conflict, and eugenics, all worldviews in which human beings—or at least elite subgroups among them—were defined in terms of their special prestige and privilege. During psychology's formative century, its dominant religion posited that our planet, along with the entire universe, was produced solely with our species in mind, and that everything in creation was delivered by divine grace exclusively for the sake of human beings. Judeo-Christian theology considered humans to reflect 'the image and likeness' of a deity (García-Alandete, 2024) and hypothesised a form of mind—the 'soul'—that literally stood apart from ordinary nature, holding a unique status within metaphysics. Classism, meanwhile, identified a human ordering within the divine hierarchy of nature, placing the wealthy and powerful on a stratum far above the masses. Eugenics, for its part, identified White Europeans and their descendants as the most important of all living beings, and decreed those of 'lesser' value to be literally dispensable. Historically speaking, the field of psychology has long been shaped by a prevailing belief in the exceptionalism of its human subjects.

Methods of Exceptionalism

Psychology's methodological practices further propel this boosting of humanity. In encouraging psychologists to examine fine details rather than big pictures, reductionism reflects the scientific tradition of applying mechanistic explanations to the conundrums of nature. But reductionism in psychology is applied selectively, focusing the field on the discrete and the quantifiable. The human mind, or 'consciousness', has always been permitted to skirt the margins of mechanisticism. Descartes's dualistic notion of a freestanding psyche, immune to biological processes, has never quite been dismissed.

Individuocentrism, for its part, leads psychology to place a permanent emphasis on human beings in isolation, rather than seeing them as part of a broader interconnected system of culture, politics, history, economics, and also *ecology*. And as we have also noted, psychology's history and methods have blended in a way that exemplifies the capitalist culture from which the field emerged—presuming humanity to be motivated by materialism, resource accumulation, and economic success. The utilitarian mind envisaged by capitalist realism is primed to view the world in terms of bounties that can be acquired, with all of nature available to be plundered in the name of self-advancement.

All these influences have shaped psychology into a field that is beholden to the specialness of human beings, creatures immersed within, but nonetheless separate from, the ecological milieu that surrounds them. Human consciousness is widely acknowledged as being a unique topic in science, ownership of which makes psychology all the more esteemed. Empirical research on neural correlates and cognitive processes is sold as addressing mysteries that might never otherwise be solved. It is commonly said that human beings are the only species capable of reflecting on their own minds in this way, a claim that simultaneously elevates the uniqueness of humanity within nature *and* that of psychology within scientific endeavour. Humans are also claimed to be the only species that authentically uses language, that has such things as 'culture' or 'civilisation', that wears clothes, produces art, reflects on its own history, ponders the distant future, or, for that matter, is aware of its own mortality. All these uniquenesses are said to define humankind. And all are embraced as the natural subject matter of psychology, rendering psychology's claims of relevance almost unassailable.

Theoretically too, psychology has developed numerous paradigms that hinge upon the supposedly singular value of humanity. Examples include humanism (the valorising of human existence), phenomenology (with its emphasis on human uniqueness), psychoanalysis (with its positing of a human-specific 'unconscious'), and cognitivism (which proclaims humankind's unique gifts for literacy, numeracy, and computational rationality). The exceptional status of humanity adds lustre to all these ideas as well as to psychology as a whole.

Psychology benefits from depicting human minds as uniquely important, *and therefore has a vested interest in doing so*. The field is fundamentally anthropocentric. It accentuates human exceptionalism by design.

The Horrors of Human Hubris

The problem here is that from the perspective of nature—that is, from the perspective of non-human species, the wider ecosystem, the climate, or the planet itself—human exceptionalism is arguably the most destructive intellectual fallacy ever to have emerged. Its many environmental consequences are well documented. Human exceptionalism has led directly to increasing global temperatures, rising sea levels, sea ice decline, worsening drought, and the overall broadening of weather extremes, as well as to ozone depletion, mass soil erosion, deforestation, reef degradation, the acidification of water systems, the proliferation of marine microplastics, and the extinctions of several species (ultimately perhaps to include that of our own). The hubris of anthropocentric exceptionalism encourages human beings to presume themselves dominant over nature in such a way as to entitle them to destroy it, without having to seriously consider how the interdependencies of their survival are corrupted by environmental neglect. Naive faith in the primacy of humankind breeds uncorroborated scepticism toward the true gravity of planetary threats, along with unwarranted optimism regarding humanity's ability to navigate, not least solve, its existential crises.

But apart from the almost tautological sense of being 'one-of-a-kind' (along with literally tens of billions of other 'one-of-a-kind' natural entities), human beings are not in any real way *exceptional*. The suggestion that humanity sits at the apex of nature's hierarchy, or represents the pinnacle of evolution, is entirely self-aggrandising. Arguably bacteria are more optimally evolved than humans, in that they seem ideally adapted for perpetual reproduction (having succeeded for billions of years so far), with enough genetic latitude to allow them to leverage natural selection to evade whatever challenges are thrown at them—collectively, for example, vanquishing the best antibiotics human scientists can design.

Human beings are part of nature as a whole and depend on nature for their survival. They rely on the natural environment for food, for water, and for air, and for everything required to maintain the much-vaunted structures of civilisation that supposedly made their species so 'exceptional' in the first place. While at times they seem intent on vandalising the planet to oblivion, destroying nature is simply not a rational option for humanity.

Human (Not) Nature

Psychology's individuocentrism and reductionism combine to promote the view that not only can human beings be studied in splendid isolation, impinged upon but separate from external forces, but also it is in fact *desirable* to think of human beings in this way. This, of course, presents a false dichotomy between *human* and *non-human*. The distinction that separates humankind from the rest of nature is more ideological assertion than statement of fact. Such *them-and-us* reasoning creates an emotional distance between humanity and nature, allowing human beings to think of the world around them as being of lesser importance than themselves, or at least as existing to serve their (human) needs.

Anthropocentrism sunders the connection between humans and the rest of the living world. It encourages a sense of trepidation, or even contempt, toward nature (Washington et al., 2021). To most of the general public, the (scientific) fact that all living species are interrelated—for example, that human beings are the genetic relatives of rodents, of insects, and even of plants—is greeted with confusion, dismissed as spiritually offensive, or grudgingly accepted as *technically* (that is to say, *not really*) true, as no more than obscure biological pedantry. Psychology's stance on the centrality of human experience serves to foster an in-group bias at the species level, echoing and thereby reinforcing a disregard for the relative welfare of non-human entities.

Psychology's assumption of capitalist realism—that there is no alternative but to frame human beings as egocentrically and materially acquisitive—is particularly destructive to nature. The history of capitalism itself can be traced back to the deforestation of Madeira in the fifteenth and sixteenth centuries, when Portuguese sugar planters stripped the north Atlantic archipelago entirely of its millions of trees, rendering the place ecologically bereft and economically worthless, and then simply moved on to do the same thing in another part of the world (a pattern of commerce occasionally referred to as 'Boom, Bust, Quit'; Monbiot & Hutchison, 2024). Capitalist realism is hubristic in that it presumes humanity to be superior. The claim that human beings can 'own' parts of the natural world, such as land or animal livestock, is itself a social construction entirely premised on humankind's self-declared pre-eminence. And yet this notion of private (human) actors owning and controlling property is an essential feature of the capitalist system.

The way the field of psychology to this day promotes human exception-alism as its core premise is entirely in tune with the ultimate needs of capitalism, by encouraging the belief that human uniqueness warrants special treatment.

But the capitalist realist idea that everything in nature should be viewed as a resource for humans—that all can be reduced to capital—is not accepted universally. It relies heavily on the consent of so-called modern industrial societies (Crist, 2020). By contrast, non-industrial cultures have historically seen humanity as cohabiters of the natural world who have little or no entitlement to property ownership (Muradian & Gomez-Beggethun, 2021). This lack of universality reveals capitalist thinking to be a byproduct of culture, rather than a funda-mental component of human minds.

The great irony of capitalist realism (and thus of mainstream psy-chology) is that materialist instincts are themselves framed as part of human 'nature', so much so that the *actual* natural world is relegated to a status subordinate to our human proclivities. Committed capitalists scoff at efforts to prioritise nature over humanity, but defend their anthropocentric biases on the grounds that human beings are 'natu-rally' exceptional. Psychology's human exceptionalism has it both ways: the exploitation of nature by humans is itself seen as part of the natural order of things.

Anthropocentrism as Conservatism

The willingness to exploit nature is far from politically neutral. Research consistently shows it to be strongly associated with hardline conservative worldviews. Human exceptionalism has been found to be strongly associated with social dominance orientation (or SDO), which in turn is a predictor of racial prejudice, sexism, homophobia, cultural nativism, and a generalised belief in the dominant position of high-status (human) groups. In this regard, human exceptionalism is linked to what social scientists refer to as 'generalised prejudice', the tendency to hold negative attitudes towards a range of different out-groups (Dhont et al., 2019).

A protectionist sentiment toward the ascendant position of humanity is balanced by discriminatory, and even oppressive, attitudes toward non-human things.

Conservatives argue that human beings are entitled to plunder nature for personal gain for no other reason than because they are human; in other words, because they are *worth it*. The field of psychology pushes many of the same buttons. It valorises humans as a species worthy of special attention in isolation from their surrounding ecosystem. It implies a dominant status for the human in-group. In so doing, it promotes the same ethos as right-wing political discourse, which pours scorn on environmentalism on the basis that the needs of humankind trump all other concerns.

As with all the topics considered in this book, it is once again important to distinguish the conventions of mainstream psychology from the political attitudes of individual psychologists. A great many psychologists are committed environmentalists, and there is ever-growing interest in examining how psychology can be used to address environmental concerns. And many psychologists are focused on addressing the adverse psychological impacts of environmental decay. There are several to consider. Environmental pollution is associated with risk for clinical depression (Khafaie et al., 2019). Chronic drought is associated with generalized anxiety and an increased incidence of suicide (Bourque & Willox, 2014). Resource scarcity is a contributor to societal disharmony (Reuveny, 2008). Escalating ambient temperatures are known to lead to increases in interpersonal violence (Anderson & DeLisi, 2011). Extreme weather events have been linked to higher levels of child abuse (Keenan et al., 2004). And environmental disruption contributes to economic collapse, with all the social determinants of mental illness that that then brings forth (Thompson, 2021).

But while so many individual psychologists are motivated to combat the climate crisis, the way their *field* depicts the world is another matter. The theoretical and methodological assumptions of psychology load in an unfortunate direction. Psychology's explanatory methods are premised on individuocentrism, reductionism, and capitalist realism, all of which feed into a promotion of human exceptionalism. And human exceptionalism is not just *linked* to conservative worldviews—it is a *manifestation* of conservatism. Psychology's promotion of human exceptionalism is therefore an inherently conservative endeavour.

A key feature of the climate crisis is that it chiefly threatens the well-being of the world's poorest and most disadvantaged populations. Conveniently, therefore, placing humans on an existential pedestal just happens to serve the egos of the world's richest and most secure humans. Meanwhile, it degrades the welfare of those whose existence is more precarious. As such, it is difficult to avoid the feeling that psychology would have ended up conceptualising human beings quite differently if individuocentrism, reductionism, and capitalist realism were found to run counter to hegemonic Western interests. If the promotion of human exceptionalism disproportionately threatened Western populations—instead of those in Africa, Asia, and the Pacific region—then we might see questions quickly raised about whether individuocentrism, reductionism, and capitalist realism are in fact the *right way* to be doing psychology after all.

As part of the juggernaut of Western culture, psychology is itself a tool of Western interests. The conveyer-belt that provides psychology's subject matter is calibrated by decision-makers who have accumulated economic and political power, raising questions as to *why* exactly some types of evidence are valued more than others, and *whose* evaluation is counted as actually mattering (Fine, 2012). Psychology's myopic focus on Western agendas runs counter to the ideals of global solidarity. In Part VI, we will examine how the West's dominance of scientific research—itself intertwined with colonialism and its many lingering commercial legacies—has led to a cultural skew in psychology, reflecting the wilful inclination of an inherently privileged community to prioritise its own needs. Psychology's unadulterated Westernness, and for that matter its Whiteness, are emblematic of an inherently conservative field—one very far removed from the supposed stereotype of excessive liberalism.

Part VI

Psychology's Whiteness Problem

34

Weird Science

Psychology, as we mentioned previously, is a science of First World problems. It is a field dominated by Euro-American concerns. Issues that affect Western sensibilities are obsessed over, while global realities are relegated. There are more studies on the psychology of school shootings than on the behavioural aspects of water hygiene, despite the fact that, each year, a million people are killed by contaminated water, whereas fewer than one hundred victims are fatally shot in schools. Of course, what makes school shootings a priority is the fact that the vast majority of school-based gun massacres occur in the United States. It is as if the psychology literature is implicitly driven by an ideology of 'America first'.

Before it declared the COVID-19 pandemic in the spring of 2020, the World Health Organization published a report on the 'grand challenges' threatening the world population's mental health (Bajbouj et al., 2022). The WHO's list included conflict-related stress, disparities in mental healthcare, poor health literacy, digital inequalities that impede access to services, and adverse effects on mental well-being caused by the global climate crisis (some of which we highlighted in Part V). But to peruse journals of psychology that year was to step into a different

B. M. Hughes, *Psychology's Quiet Conservatism*,
https://doi.org/10.1007/978-3-032-07724-0_34

reality. The top three articles downloaded from the major psychology journals in 2020 comprised a paper on 'selfie behaviours' in young women (the article was titled *Me, My Selfie, and I*), a study of high-school students' perceptions of their physical appearance on social media, and a commentary discussing the impact of news reporting on anxiety aroused by so-called 'collective crises' (Palmer, 2021). It seems that psychology's own anxieties did not extend as far as the world's grandest challenges.

That most psychology research is conducted in the United States is widely acknowledged. But even that observation understates the narrowness of the field's coverage. Geography fails to account for the selectivity of psychology's purview. For example, two-thirds of participants in the studies featured in the *Journal of Personality and Social Psychology* are not merely American, they are also college students—and not just any old college students, but students enrolled in courses in psychology (Hughes, 2023). Hence the oft-quoted quip that the long-esteemed outlet should rename itself the *Journal of Personality and Social Psychology of American Undergraduate Psychology Students* (Henrich et al., 2010).

College students taking psychology degrees are an especially niche group, heavily stratified by socioeconomic status. But even for non-college populations, participating in psychology research almost always requires some level of affluence. Turning up at research laboratories or filling out surveys takes time. And time is money. Participating in research is an activity that people at the margins of society are unlikely to prioritise. Some types of research are especially skewed as a result. For example, most research in child psychology relies on a small subset of families who have the resources, awareness, and motivation to enrol their offspring into a study. As a consequence, journals of developmental psychology typically feature samples that are even *less* diverse than those appearing in the ordinary journals of American undergraduate psychology students (Fernald, 2010).

With most of its research conducted on affluent Americans who have (or whose parents have) contact with higher education, it is little wonder that psychology has so far accumulated limited evidence on malnourishment, malaria, or the migrant experience. Notably, research that takes place *outside* the United States is skewed in essentially the same fashion. Overall, some 96% of research samples in psychology are composed of people who live in North America, Europe, Australia, or Israel, a set of regions with overlapping cultures that together represents just twelve per cent of the world's population. As noted in Part II, this group of territories is nowadays commonly described as 'WEIRD'—standing for Western, Educated, Industrialised, Rich, and Democratic—an acronym proposed by University of British Columbia psychologists Joseph Henrich, Steven Heine, and Ara Norenzayan (Henrich et al., 2010). With the proliferation of internet-based survey studies in psychology over the past decade, some observers suggest that we can now add an 'O', standing for *Online*, to the label. Psychology is thereby cast as the official science of WEIRDOs (Colvile, 2016).

The core problem here is less one of democratic representation than one of scientific thoroughness. Psychology has pretensions to be the 'science of mental life', but in pursuing this status, it has limited itself to the mental lives of a niche subset of humankind that happens to live in rarefied environments (Hughes, 2018a). Nobody would accept marine biology as the 'science of aquatic life' if it only ever chose to study, say, farmed lobsters bred in the North Atlantic.

Some psychologists have attempted to rationalise their field's WEIRD bias by claiming that diversity, per se, isn't what it used to be. They point out that cultural colonialism is gradually blending the entire planet into a single homogenous society. In other words, psychology is fine the way it is, because soon enough *all* humans will end up WEIRD anyway (Cooperrider, 2019). A related suggestion is that, despite making up just one eighth of humanity, people in WEIRD societies are in fact the perfect example of what human beings are 'really' like. Their lives are unencumbered by the many psychologically distorting impositions that the majority-world population are forced to bear out of necessity (Maryanski, 2010). This overall argument—that all human societies naturally gravitate toward WEIRDness—rests on the implication that WEIRD environments are inherently attractive to humankind, because they allow their inhabitants the freedom to be *themselves*, a freedom that (supposedly) does not exist in traditionalist, collectivist, impoverished, or undemocratic places.

As mentioned throughout this book, the view that human beings are naturally capitalistic—the central claim of capitalist realism that is shared by mainstream psychology—is itself highly arbitrary. But notably, it enjoys disproportionate political support in precisely those parts of the world with which psychology has become obsessed. In other words, capitalist realism is a WEIRD idea that primarily resonates with WEIRD-minded people. Its currency within psychology reflects the very fact that psychology is subject to the echo chamber of WEIRD bias. This itself reflects the distorting influence of global economic power on international academia. Virtually all of psychology's textbooks and journals are produced within the highly capitalistic United States, or in that country's cultural orbit (a constituency that includes the UK and, at a slightly more diluted remove, the European Union). It is little wonder that psychology takes the capitalist worldview as its epistemic starting point.

The economic pre-eminence enjoyed by the US, the UK, and European countries has itself emerged in large part as a consequence of a number of problematic historical forces, including imperialism, colonialism, and cultural elitism. This history brings a wider set of issues to the fore. Because to worry about 'WEIRD psychology' is to avoid the fact that the WEIRD label is not merely a synonym for 'Western' per se. It is also, in essence, a euphemism for something else. The term misleads us into thinking that the core issues are geography and technology, when in fact they are political and historical. The so-called WEIRD bias conceals deeper human schisms.

Human identities are intersectional but there are many common conflations. For example, a large majority of the WEIRD population belongs to a single ethnic category. Generally speaking, the other people who live in WEIRD places find themselves marginalised in terms of economy, well-being, and power. Therefore, in itself, the distinction between WEIRD and non-WEIRD represents a false dichotomy, because it ignores the fact that skews exist *within* the WEIRD world as well as beyond it. Not everyone who lives in a WEIRD society is served well by psychology—because they are not served well *by that very society.*

Psychology's WEIRD skew conceals—but is propelled by—a transcending category of bias that is not even referred to in its five-part acronym, one that is possibly more important than those that *are* highlighted. That psychology is the science of WEIRDOs is a deflection. It is more important to state that, in many respects, psychology is a science of, and for, White people.

35

Structural Racism in Psychology

Psychology's problematic relationship with Whiteness[1] is reflected in the demographics of its workforce in WEIRD countries. In the United States, nearly 90% of psychologists who work in the health services are White, compared to just 60% of the general population. Meanwhile, just 3% of health service psychologists are Black, despite the fact that Black people make up 12% of American society (Lin et al., 2018). In other words, there are one-and-a-half times more White psychologists in the US than there should be based on

---

[1] Terms referring to ethnicity are socially constructed, have evolved historically, and are frequently contested. As an author, I completely acknowledge my own positionality as a (so-called) White person who lives in a White-majority society. On these pages I use terms that are themselves embedded in the very controversies under discussion—terms such as 'Black', 'White', 'Whiteness', 'race', 'ethnic', 'minorities', and so on—but I do so in order that readers can relate these words to the problematic discourses that I am attempting to critique. Unlike some other writers, I choose to capitalise the word 'White', as a way of highlighting its rootedness in the discourse of White supremacy. As discussed in Part II, the very act of classifying human beings into ethnic groups using fixed labels is scientifically and conceptually flawed. Moreover, terms such as 'Black' are themselves often reductionist projections of White-centredness, and so can be thoroughly problematic. My overall purpose here is to employ terminology only insofar as it refers to socially constructed and politically-motivated labelling practices—hoping, in part, to highlight the damage such practices cause—and not to imply a corresponding structure inherent in the human population.

© The Author(s), under exclusive license to Springer Nature Switzerland AG 2025
B. M. Hughes, *Psychology's Quiet Conservatism*,
https://doi.org/10.1007/978-3-032-07724-0_35

population demographics—and just a *quarter* the number of Black psychologists. Similar problems affect other branches of psychology: for example, over 91% of educational psychologists are White (US Bureau of Labor Statistics, 2024). And White people with doctorates in psychology are much more likely than their Black counterparts to secure jobs in academia: over 80% of psychology faculty are White, compared to around 70% of doctorate holders overall (Lin et al., 2018).

The imbalance represented by these figures is significant, and extends beyond the general educational inequities faced by non-White Americans. Simply put, in the US, progressing through psychology's professional pathways just seems easier if you are White. This is despite the fact that the field of psychology, and its practitioners, are said to be excessively *liberal*.

Similar patterns of White advantage are frequently found in Europe. According to the UK census, some 83% of the British population are recorded as White, nearly nine per cent are recorded as Asian, and around 4% are recorded as Black (Office for National Statistics, 2023). Meanwhile, 90% of registered practitioner psychologists in the UK self-identify as White, 4.3% identify as Asian, and just 1.7% identify as Black (Kanceljak & Calia, 2023). While the skew is not as stark as in the US, the trend by which White people are more represented than non-White people is nonetheless discernible.

Once again there is a funnelling effect. In the application system for clinical psychology doctorate programmes in British universities, White applicants are statistically more likely than Asian or Black applicants to obtain one of the highly coveted government-funded training places. White candidates submit a majority of the applications (72.6%) but are awarded an even larger majority of the places (75.5%). Meanwhile, Asian and Black candidates, while together making up 16.2% of the applications, receive just 13.4% of the places (Clearing House for Postgraduate Courses in Clinical Psychology, 2024). In other words, despite ostensibly meeting the same academic entry requirements, non-White psychology graduates appear much less likely to gain admission to postgraduate courses than their White classmates. There would seem to be a glass ceiling for ethnic minorities in British psychology.

The glass ceiling largely results from the ancillary criteria that applicants must meet to be competitive in the selection process. For example, applicants who complete *additional* academic courses—beyond their basic psychology degree—are much more likely to attain a place than applicants who hold just one qualification (even though just one qualification is 'required'). Applicants who have published a paper in an academic journal are much more likely to succeed than applicants who have not done so (even though publication is not 'required'). Applicants who have accumulated years of post-study work experience in a psychology-relevant setting will be much more likely to be admitted to clinical training than applicants who are just out of college (even though work experience is not 'required'). And so on, and so on.

In short, entry to clinical psychology training is reserved primarily for applicants who are able to access additional, rarefied, and expensive occupational opportunities prior to making their applications. Not everyone can afford to complete discretionary additional university degrees. Not everyone has the social capital, or economic means, with which to spend years working in voluntary or low-paid placements. These experiences are not professional qualifications, they are socio-economic differentiators. Simply put, given the pre-existing inequalities in UK society, it is no surprise that a system that chooses to reward such arbitrary criteria regularly proves easier for White people to navigate.

Psychology's White-centred Professional Cultures

Professional psychology's glass ceiling is further maintained by more direct forms of racial prejudice. A notable example of this occurred at the 2020 annual conference of the British Psychological Society's Division of Clinical Psychology, which was held in Solihull, England. Dr Kimberly Sham Ku, a forensic psychologist from Birmingham, had planned to present a poster describing her work supporting African and Caribbean men in forensic services. After arriving at the exhibition space, she was shocked to discover that someone had defaced her work with graffiti. The slogan 'Keep BME [Black and minority ethnic persons] out of services' had been scrawled across her poster. The conference was a private event, which meant that only other conference delegates—that is to say, *only other psychologists*—had access to the room. But when Dr Sham Ku complained to the organisers, they simply offered to reimburse her printing costs, something she felt was of utterly negligible value (Sham Ku, 2020).

Some months later, the BPS did publish an apology to Dr Sham Ku. They also announced that it had 'informed' the person who defaced the poster that 'they are now banned from attending any future BPS events' (Bajwa, 2020a). It is something of a pathetic irony that the Solihull conference was intended to highlight the specific issue of racism and social inequality in British psychology. Given the enormous publicity generated by the racial intimidation of Dr Sham Ku, you could say that it succeeded in doing precisely that.

In recent years, professional psychology in the UK has found itself embroiled in a number of racism controversies. Many have involved overt interpersonal harassment, such as that endured by Dr Sham Ku. Trainee psychologists have encountered racism so frequently that, in 2019, the British journal *Clinical Psychology Forum* dedicated an entire special issue to the matter (Wood & Patel, 2019). But other controversies have reflected a broader set of problems relating to race. They convey a sense that British psychology is unthinkingly centred around the Whiteness of the country's majority population—and insensitive toward, if not dismissive of, the lived experiences of psychologists (and clients) who are of African, Caribbean, Asian, or other non-White or colonial heritage.

Perhaps the starkest example of this set of issues took place at another BPS event, the annual conference of the BPS Group of Trainers in Clinical Psychology, held in Liverpool in 2019. As part of the conference's customary social programme, delegates were hosted at a gala dinner in the venue's banquet hall, complete with dimmed lighting, sparkling wine, and music. As dessert was served, a group of dancers emerged to perform Capoeira, a traditional Angolan-Brazilian martial art based around dance and acrobatics. The conference was being held at a venue adjacent to Liverpool's International Slavery Museum, so for the occasion the dancers performed a 12-minute dramatisation based on the transatlantic slave trade. Specifically, they re-enacted a slave auction, parading 'slaves' around the banquet hall, pointing out that one 'slave' had all his teeth and praising 'the strength of his legs', and inviting members of the audience to 'bid' for a slave of their own, all while the BPS delegates finished their tiramisus.

Many of the attendees were immediately appalled at this spectacle. Most who were Black or of other non-White heritage left the room, several visibly enraged, others crying. But the majority of the psychologists—White people, in the main—simply stayed behind, apparently enjoying what they saw, applauding the show, and even joining in with the dancing. They then stuck around to participate in a slavery-themed table quiz over late-night drinks, their Black colleagues having retreated in tears (Patel et al., 2019).

When challenged, the conference organisers later issued a lengthy apology, expressing horror at the fact that people had been offended. They admitted that they had not anticipated the 'overall emotional tone' of the act they had booked for that evening; however, they also pointed out that the Capoeira performance was intended to be educational rather than entertaining (Patel et al., 2019). If anything, such a caveated apology only made things worse. It emphasised the White-centredness of the organisers' worldview. It failed to acknowledge that re-enacting a slave auction would be traumatising for delegates whose own ancestors had been brutalised by slavery. It ignored the fact that few Black psychologists needed to be 'educated' on the history of slavery. And most of all, the apology was just not issued quickly. It emerged only after what critics describe as the usual period of 'white silence' (Wood, 2020).

Psychology's White-centred Professional Histories

As we saw in Part II, the eugenics movement bequeathed to psychology its primary statistical and psychometric practices, as well as a number of highly questionable theoretical notions about human hierarchy and heredity. For decades, advocates and apologists of the eugenic approach were heralded as paragons of British psychology, among them Galton, Pearson, Fisher, Burt, Spearman, Eysenck, and Cattell. The fact that these particular White men were placed on a professional pedestal remains indelible within the history of the field in the UK.

In 2020, responding to the ingrained Whiteness of the UK profession, the chief executive of the British Psychological Society, himself of British-Asian heritage, declared his organisation to be 'institutionally racist' (Bajwa, 2020b). His pronouncement coincided with the BPS's promotion of a task force dedicated to diversity and inclusion. *The Psychologist,* the society's monthly magazine, reported these developments in a special anti-racism issue, which solemnly bore an all-black front cover (Sutton, 2020). But while these initiatives were widely well received, several BPS members took a different view and made formal complaints to the organisation. They condemned their professional body for its 'social justice' agenda, which they felt was an inherent failing, and bemoaned what they claimed was the BPS's increasing 'left wing bias' (Wood, 2020). It appears that British psychology is no more immune than any other part of modern society to culture-war conservatism, and its backlash against diversity and inclusion initiatives.

For its part, the US profession has had its own fraught history. We have already discussed how many of the American Psychological Association's early presidents were committed eugenicists and race supremacists. In the twentieth century, several prominent American psychologists were vocal in opposing immigration and in defending racial segregation. Most of their colleagues opted to remain silent rather than comment on any bigotry within their field. We noted how in the 1960s, the APA discouraged its members from voicing support for civil rights, ostensibly because of concerns about professional neutrality, but nonetheless consistent with the political sensibilities of what was then a majority-White US population, within which strident anti-Black racism had become endemic.

As detailed in Part III, American mental health clinicians have long sought to pathologise objections to structural racism, in the past using diagnoses such as 'drapetomania' and 'protest psychosis'. American psychology also mainstreamed the practice of attributing problematic mental states to individuals rather than to cultural or historical factors, such as when linking police violence to racist officers rather than to internalised White-centric, post-slavery social norms that legitimise the brutalising of Black people. And, as we saw in Part IV, professional psychology—largely (but not only) in the US—developed and promoted the use of psychometric instruments known to have inherent cultural biases that disadvantage non-White people, with the APA frequently being criticised for failing to act to end the misuse of skewed testing practices (Gómez, Caño, & Baltes, 2021).

American psychology has also been intertwined with several US-specific racism controversies. These include psychologists' decades-long service in the so-called American Indian boarding schools system, where Indigenous youths were forced to speak English and prevented from expressing their traditional cultures (Cummings Center, 2021). US psychologists have been a key part of a mental healthcare infrastructure in which Black patients are disproportionately misdiagnosed and more likely to be subject to involuntary psychiatric confinement (Hicks, 2004). Indeed, the APA continues to provide training accreditation to several hospitals that demonstrably overdiagnose Black patients (Auguste et al., 2021). Research has shown that the therapies provided by American psychologists are largely White-centred, rarely taking account of clients' cultural or linguistic backgrounds (Sue & Zane, 2006) and frequently conveying microaggressions toward non-White people (Constantine, 2007). A lasting consequence of such shortcomings is that many non-White Americans are markedly reluctant to seek mental healthcare, a factor that compounds already troubling levels of health marginalisation and outcome inequality.

In 2021, the APA issued a lengthy statement apologising for its contribution to systemic racism in the US over the previous century-and-a-half (APA, 2021). However, the document focused mostly on historical policies of the APA itself, and its support for practices such as eugenics and scientific racism. Critics accused the APA of 'cherry-pick [ing] its way through history' by failing to address psychology's role in suppressing political dissent (Auguste et al., 2021) and in pathologising resistance to racism (ABPsi, 2021). The statement did not consider the field's broader involvement in preserving society's imbalanced power structures (for example, when promoting individuocentrism ahead of more sociopolitically informed analyses). And as it focused only on events in the United States, the apology was guilty of ignoring the implications of WEIRD bias and its promotion of an implicitly White-centred worldview among psychologists. The American Association of Black Psychologists (known as ABPsi) issued its own statement rejecting the APA's apology outright. In its view, the apology was a superficial gesture by the APA to rid itself of 'white guilt'. It was 'at best patronizing,' the ABPsi statement observed, 'and at worst, an intentional act of obfuscation designed to mask the truth' (ABPsi, 2021).

Ultimately, the professional structures of psychology provide little evidence that the field, or its practitioners, are in thrall to some all-consuming liberal bias. Instead they reveal a profession of psychology that is similar to just about all other institutions and organisations in the Western world—disproportionately populated by White people, who, while frequently voicing distress at the spectre of racism in their midst, nonetheless seem to hold all the important positions, and always manage to gatekeep their guild in a way that just happens to perpetuate their own arbitrarily elevated status.

36

Mechanisms of Whiteness

Many professions face challenges with representation and inclusion. However, for psychology, a field that purports to offer a science of human perspectives, the consequences are especially problematic. It is widely understood that psychological interventions work best when client and practitioner are able to identify with each other, when they can share a common perspective on life upon which progress can be built. But such alignment is difficult to achieve in Western countries, where it is mostly White practitioners who are available to support people with their psychological needs. This is particularly ironic as many such needs stem from forms of marginalisation that disproportionately affect non-White people. Given the psychology profession's ethnic homogeneity, opportunities for effective interventions with non-White people—whether in clinical care, education, or industry—are compromised from the get-go.

Even the way psychology builds its scientific knowledge base is, for all intents and purposes, disproportionately 'White'. Because of its WEIRDness, psychology draws most of its data from societies that are dominated by White people, White histories, and White concerns. A number of legacy issues compound this pro-Whiteness bias. Firstly, psychology defaults to Eurocentric ideas about human motivation that are premised on the materialistic economies of White-centred societies. Secondly, psychology carries the vestiges of imperial colonialism not only in its use of (White) Europeans as its template for interpreting human minds, but also in its systems for prioritising, producing, and promulgating evidence. This so-called 'global' discipline's academic infrastructures were built for the benefit of White-majority motherlands.

And thirdly, psychology exhibits a form of social organisation that confuses social-justice *language* with social justice itself, blind to the fact that its practitioners and proponents have a vested interest in preserving the societal status quo along with all its imperfections. Psychology's so-called liberal tendencies have been appropriated by a stratum of proxy advocates to whom the authentically marginalised are expected to defer—psychologists who are well-meaning but privilege-blind, who eloquently talk the 'social justice' talk but who walk in a very different direction.

We will now examine each of these legacy issues in turn.

Eurocentric Materialism

The term 'White supremacism' is widely considered to be incendiary, a slur reserved for describing only the most extreme manifestations of racial prejudice. The phrase acquired this connotation through several centuries in which the ruling regimes of Europe—and later America—saw fit to pursue for themselves cultural and economic dominance over the rest of the world, invoking notions of their own pre-eminence to justify the use of coercion, subjugation, and military force. Historically, White supremacism has underpinned such contemptible horrors as slavery, imperialist conquest, and genocide. It has also inspired several racial-segregation and apartheid regimes. And as an ideology it continues to energise extremist political movements in Europe and North America, spawning such outlandish racist claims as the 'White replacement' (or 'White genocide') conspiracy theory, and finding expression—along with some electoral success—in anti-immigration and anti-DEI policies as espoused by ethno-nationalists, neo-fascists, and Far Right political groups.

Few liberal psychologists would consider themselves 'White supremacists' in *any* possible sense of the term. And yet, when taken literally, the idea of White supremacism implies an inherent belief in the exceptional nature of White people—and, by extension, of those cultures and countries that can reasonably be identified as historically 'White' domains. As outlined above, the field of psychology adopts a predominantly WEIRD perspective when constructing its understanding of human behaviour. It presumes that 12% of the human population provides an ample basis on which to draw conclusions about all others, and prioritises the needs of the US and Europe when choosing whose problems require attention. Psychology's in-built focus on White-majority countries looks an awful lot like something that presumes these places to be of inherent importance to the world. It is a view shared by those who feel that White *people* are innately superior. The line between extremist political ideology and everyday in-group favouritism can sometimes be difficult to place.

The bias in psychology is more implicit than explicit. Recall the claim that WEIRD societies provide a uniquely valid perspective on human nature because they liberate people to *be themselves*, unencumbered by traditionalism, collectivism, or poverty (Maryanski, 2010). The implication is that for a human habitat to be WEIRD is an inherently good thing: it provides conditions that are 'adaptive' in evolutionary terms. But the framing of WEIRD societies as 'free' depicts non-WEIRD societies as constrained. The elements of non-capitalist bureaucracy, such as social safety nets, are portrayed as cultural artefacts that stifle human self-expression. This is the argument that states why capitalism is dominant over socialism: it valorises self-interest and competition, instead of clinging to collectivism. Notably, the unspoken thrust running through this entire case is that it was White people who built these supposedly superior systems in the first place.

But the implication that non-WEIRD values are being driven to global extinction by hegemonic Western 'progress' is less scientific observation, and more political persuasion. Survival-of-the-fittest claims that Western culture must be desirable simply because it spreads at the expense of others make sense only if you take the view that evolutionary triumphs are imbued with a moral valence. However, natural outcomes are not guided by morality. They provide no insight on what is morally worthwhile for human beings. The bubonic plague, for example, once threatened to take over the world, without exactly enhancing the plight of humanity. Humans die from 'natural' causes all the time without ever *deserving* to do so. The idea that natural processes inevitably reinforce that which is best for humanity might be comforting to some people in the present moment, but, as with all forms of just-world thinking, it is based on little more than wish-fulfilment. There is nothing other than prejudice to corroborate the idea that WEIRD societies are somehow more 'liberating' for human nature.

In fact, there is plenty of evidence (much of it from WEIRD research) to show that capitalist thinking is detrimental to human minds. People who place a heavy emphasis on materialism—on ensuring wealth and gathering possessions, and on cultivating image and status—report diminished, rather than enhanced, mental well-being (Dittmar et al., 2014). They experience more negative self-appraisals (Richins, 2004), lower life satisfaction (Garðarsdóttir et al., 2009), greater negative affect (Kasser et al., 2014), and engage in more risky health behaviours (Auerbach et al., 2010). Pursuing materialist goals is associated with higher levels of burnout and lower career satisfaction (Deckop et al., 2010). People who score high for materialism experience lower-quality interpersonal relationships (Solberg et al., 2004), more loneliness (Pieters, 2013), and more anxious attachment styles (Norris et al., 2012). They also score lower for personality traits that ordinarily underpin happy lives, such as empathy (Sheldon & Kasser, 1995). They are more likely to engage in antisocial behaviour (Deckop et al., 2015), less likely to care about social responsibilities (Kolodinsky et al., 2010), and even have larger carbon footprints (Kasser, 2016). Children who express materialistic ambitions report poor intrinsic motivation and get lower grades at school (Ku et al., 2014).

And as mentioned previously, contrary to the assumptions of capitalist realism, there are in fact many alternative economic philosophies that, simply by virtue of their existence, show how human minds are quite capable of thinking differently about material wealth. Such alternative worldviews have long been sidelined by Western psychology, with their contributions airbrushed from its mainstream histories.

Perhaps the most conspicuous examples are Marxist theories about the role of motivation in human nature. Dialectical materialism, the philosophy that underpins communism, is one such psychological theory. It emphasises how the innate human capacity for 'conscious purposeful action' exerts a greater influence on human behaviour than personality types, group dynamics, or operant conditioning (Holowinsky, 1985). An important implication of dialectical materialism is to suggest that human psychology is naturally intertwined with systems of social organisation. Another is that intellectual pursuits such as religion, art, and culture, emerge when the psychological experiences of the masses percolate upwards through society, rather than when something is transmitted downward by those who occupy civilisation's 'higher' echelons.

In essence, dialectical materialism offers a comprehensive challenge to psychology's individuocentrism and reductionism. Its ideas are eminently reasonable to consider, but yet—largely because of the Cold War—have disappeared from mainstream psychology. The Russian neuropsychologist Alexander Luria was a prominent advocate of the paradigm, but, as noted earlier, modern psychology textbooks rarely mention his work in this area (Kotik-Friedgut & Ardila, 2020). Instead, as far as today's psychology students are concerned, Luria is to be acclaimed only for his research on brain signalling, and not for his views on the interactions between brain development and political history.

Psychology's focus on WEIRD populations is intertwined with its embrace of capitalist realism as the universal template for understanding human minds. It reflects the in-group bias of Euro-American psychologists who play a dominant role in the supply chains of academia. In part, this dominant position itself reflects a history of colonialism that has shaped the modern world in which we now live. Having built their status on the exploitation of other places, WEIRD countries today remain disproportionately privileged. The fact that these countries are majority-'White' is inextricably intertwined with how colonialism works.

Colonialist Psychology

Fundamentally, colonialism refers to the practice where powerful actors coercively acquire control over a foreign place or people, with the aim of exploiting them to serve their own interests. It also involves the imposition of the colonisers' own ways and norms onto the communities that are colonised, and the eradication—or at least the disregard—of those people's native customs. Most discussions about colonialism are framed in terms of political history, where one country seeks to colonise a foreign territory through military conquest, or where settlers invade and seize foreign lands for themselves and drive away, or subjugate, the people who lived there before them.

Psychology's history spans both colonisers and colonised peoples. But the very result of colonialism is that the coloniser's version ultimately takes precedence. The problem for psychology, of course, is that the experiences of rich Europeans and Americans offer a poor basis from which to draw conclusions about people elsewhere in the world.

When attempting to apply Western psychotherapeutic principles to his patients in North Africa, the psychiatrist and activist Frantz Fanon famously found that 'Freud's discoveries are of no use to us' ('*les découvertes de Freud ne nous sont d'aucune utilité*'; Fanon, 1952; p. 84). Fanon was born in the French colony of Martinique, served in the French army during World War II, and later worked as a physician in Algeria. He offered several profound insights into the colonial nature of psychology and the inapplicability of psychological ideas outside the narrow contexts in which they are produced. As well as the limitation that most psychology is framed from a Euro-American (and thus predominantly 'White') perspective, Fanon showed how the colonial force of psychology's 'White gaze'—its '*regard blanc*' (p. 89)—is directly harmful to Black and other non-White people. His work drew attention to the intricate ways in which psychological concepts interact with social and political contexts, and how psychological experiences cannot be understood without taking account of the conditions within which a person lives (Fanon, 1961).

One example is trauma, which Fanon encountered frequently during wartime (Goozee, 2021). Psychology's approach is to focus on the *consequences* of trauma—on the symptoms of *post*-traumatic stress—rather than on its causes, which include all the political events, such as war or oppression, that lead people to become traumatised in the first place. As such, according to Fanon, standard psychology seems more interested in reintegrating trauma patients back into a trauma-inducing world, rather than in doing anything to change that world for the better or to hold traumatisers to account. Fanon's assessment that psychology reifies exploitative power structures, and so is unassailably political, should have been a landmark contribution to the discipline. However, few psychology students are ever taught about Fanon's ideas. It is as though his (non-White) face simply doesn't fit the mainstream psychology curriculum. Fanon's inspirational challenge to colonialist psychology has itself been pushed to the periphery.

Similar plights have befallen other efforts to re-configure psychology beyond its colonialist form. Inspired by working with illiterate and impoverished communities in Brazil in the mid-twentieth century, educational theorist Paulo Freire argued that traditional models of psychology encouraged passivity in learners by treating them as vessels to be filled with information, rather than as co-creators of knowledge. His own approach, critical pedagogy, emphasised dialogue and critical thinking within social and political contexts (Freire, 1970). However, although influential across the social sciences, Freire's work has been marginalised by psychology, where it is typically regarded as unduly focused on political dynamics at the expense of cognitive processes.

Likewise, the ideas of Ignacio Martín-Baró, a priest and psychologist who worked in El Salvador, are rarely discussed in mainstream psychology textbooks. Martín-Baró established an approach to social psychology that was specifically framed around the contexts of oppressive sociopolitical structures, such as those he witnessed in Latin America in the 1970s. His ideas led to the development of a subfield now known as 'liberation psychology' (Gondra, 2013). However, despite its questioning of mainstream psychological theories, Martín-Baró's work remains at the fringes of the field.

While it is true that critiques such as those of Fanon, Freire, and Martín-Baró garner academic attention today, it is only insofar as they are understood to represent views from the margins of psychological scholarship. They are seen as niche ideas unconnected to the 'formal' or 'orthodox' science of psychology, essentially lying outside the field's colonialist remit.

Colonialism, per se, need not involve state actors. In the modern world it can sometimes take the form of cultural hegemony—where wealth, media influence, and social capital are deployed to preserve a single elite culture's self-rewarding dominance. Psychology's WEIRD bias reflects this cultural version of colonialism too. The academic field is dominated by researchers at American and European universities, who owe their influence to the good fortune of being located in countries that have been enriched by past colonial exploits. More than that, psychology preserves its monocultural identity by exerting influence through the infrastructure of modern academia, itself an edifice shaped by colonial history.

The major grant-awarding agencies, universities, academic publishers, and literature databases are all headquartered deep inside the WEIRD bubble. All operate according to its particular market demands. The leading journals and textbooks are published in Europe and North America, mainly by private corporations. And they are only designated as 'leading' on the basis of bibliometric indices, such as journal impact factors, that they themselves devised for (self-)promotional purposes. Similarly, academic search engines are the inventions of Western commercial companies, and are thus calibrated for Western audiences. It is consequently easier to locate psychology studies on 'selfies' than research on 'saving face' or 'caste systems', because the latter two topics, while hugely influential in the lives of billions of people, relate to culturally non-Western concerns.

Academics in psychology throughout the world are expected to use this colonial infrastructure to keep abreast of their field. American or European academics will rarely be criticised for ignoring Chinese, Indian, or African psychology journals. But scholars from Chinese, Indian, and African universities will be considered unemployable if they fail to keep up to date with the key American and European outlets

(Pickren, 2009)—including, no doubt, the *Journal of Personality and Social Psychology of American Undergraduate Psychology Students*. In addition, researchers outside the Western bubble are rewarded for publishing inside the bubble, but the reverse is rarely ever true. What results is a virtual brain drain. Academics in South America, Asia, and Africa are incentivised to spend their career-time conducting studies that can compete for space in journals whose aims and objectives are geared toward North American and European needs (Alatas, 2003).

The problems produced by all these double-standards are not confined to academia. For example, a key debate early in the COVID-19 pandemic was whether it was reasonable, or even safe, to rely on behavioural interventions derived from WEIRD-focused research when working with diverse global communities (IJzerman et al., 2020). It is similarly questionable whether procedures developed in the US or Europe can accurately diagnose conditions such as ADHD in children who, for example, live in rural villages in India (Sridhar, 2024). And Western psychological tools, such as personality tests and motivational techniques, are unlikely to resonate with workforces in places where traditional community outlooks are less compatible with capitalism (Bhatia & Priya, 2018).

It is worth stating again that the WEIRD sphere is composed—entirely—of countries with White-majority populations. So the scenario that eventuates from academic colonialism is one where *non*-White academics are expected to conform to the standards of *White*-majority cultures. This is why psychology's WEIRD skew is an inextricably racial bias. It demands that non-White psychologists around the world professionally submit to the gravitational pull of WEIRDness.

Elite Capture

As noted repeatedly throughout this book, when it comes to considering the impact of psychology as a field—that of its methods, practices, and teachings—the personal views of individual psychologists are neither here nor there. Psychology can preserve various forms of systemic privilege regardless of the (supposedly) progressive politics of its practitioners. Individual psychologists might well profess themselves to be 'liberal', or even 'woke'. But the epistemological instruments of psychology—its individuocentrism, its reductionism, its quantification practices, its inherited theories and assumptions, its embrace of capitalist realism, and its colonial (i.e., White-centric) biases—combine to reinforce humanity's harshest power imbalances, thereby fulfilling a thoroughly conservative agenda.

In the main, psychologists are bystanders to, and beneficiaries of, the status quo. The current standards place psychologists on a pedestal from which few are inclined to step off. Psychologists' immersion in social prestige begins early in life. Most come from socioeconomically prosperous (or at least undeprived) backgrounds. When attending college, they receive the compounding benefits of higher education, a system from which economically marginalised and socially disenfranchised communities are customarily excluded. The choice to embark on a career in psychology itself garners further prestige. Psychology is widely regarded to be a high-status occupation, as is evidenced by the high demand for places on university psychology courses (in the UK, for example, psychology is the country's most popular science subject and its second-most applied-for degree overall; Rogers, 2025). Qualified psychologists are seen as socially privileged professionals, with most carrying titles, such as 'Doctor', that publicly signify their elevated status. And they typically go on to earn substantial salaries, once the neoliberal defects of precarious employment have been navigated.

Psychologists also benefit from more indirect attributions of esteem. The very work that psychologists do is usually regarded as being inherently beneficial to society. The services psychologists purvey are understood to be evidence-based, and are therefore perceived as being especially valuable, conferring humanity with the enhancements of empirical science. More superficial prejudices also creep into consideration: most psychologists in Europe and North America—psychology's 'centre' as opposed to its 'periphery'—are White, which by dint of colonial history, is widely regarded (often unconsciously) as signifying authority, ascendency, and academic seniority.

So, why would the members of a supposedly 'liberally biased' psychology stand by and allow this edifice of privilege to remain intact? The reality, of course, is that as the majority of psychologists benefit from existing arrangements, few are keen to see those arrangements change. It is simply human nature to suppress survivor's guilt. Psychologists may vote for liberal politicians, support liberal causes, and subscribe to liberal newspapers, but this is more for tribal reasons than because of an authentic belief that *they themselves* enjoy undeserved privilege and need to be brought down a notch or two. Or to put it more bluntly: the field of psychology is over-populated by rich White people who see themselves as committed to social justice, but who rarely work toward it in a way that actually costs them anything.

Claims of 'liberal bias' are a form of identity politics. Right-wing commentators deploy identity politics in a divide-and-conquer fashion, to deflect attention away from those who enjoy undeserved privilege, and to scapegoat, through ridicule, the woke and politically correct, whose sentiments can be blamed for all of life's ills. But those on the left also take advantage of identity politics, albeit in a different manner. Typically, the left employ the politics of identity as a virtuous cause, with which to signal their own integrity without having to accommodate actual reform. It is this form of identity politics that American philosopher Olúfẹ́mi O. Táíwò describes as exemplifying 'elite capture' (Táíwò, 2022).

For Táíwò, supposedly progressive institutions and agendas are regularly hijacked by advantaged actors, who are motivated to preserve the status quo while appearing to challenge it (but while genuinely believing themselves to be standing up for the little guy). The strong present themselves as acting in the interests of the weak—but they decline to allow the weak to truly act for themselves, *for fear that they might actually achieve their own emancipation.* One feature of elite capture is a type of performative deference, in which the marginalised are afforded a superficial platform within activism, but only on terms that suit those who benefit from their marginalisation. Táíwò points out that such norms of deference 'provide social cover for the abdication of responsibility' (p. 81). This is why the privileged classes so frequently direct attention toward symbolic causes, such as the selection of spokespersons or boardroom representation, and away from root political issues, such as structural racism. In many respects, the modus operandi of Táíwò's elite capture reflects psychology's use of individuocentrism—efforts are focused on individual perspectives rather than on wholesale systemic reform.

In reality, mainstream psychology conforms to the needs of those who already enjoy privilege in Western society. Its progressive stances are curtailed by a residual loyalty to the already prioritised. As a result, despite appearances, much of what goes on in psychology operates directly against inclusion, diversity, and global solidarity.

We have already seen how the APA's attempt at an apology for racism fell dramatically short. It was duly condemned by actual victims of racism for focusing on symbolic causes with selective hindsight, rather than doing anything that might threaten today's embedded power imbalances. As we saw in Part V, psychology's discourse on racism is frequently focused on interpersonal prejudices, promoting the use of one-to-one anti-racism interventions to rehabilitate prejudiced individuals. Its individuocentrism and reductionism ensure that efforts to invoke structural, institutional, or colonialist contexts are displaced, in a process South African psychologist Kevin Durrheim identifies as 'conversational silencing' (Durrheim, 2024)—talking about Topic A as a way of *not* talking about Topic B.

Psychology succumbs to elite capture when it speaks out on racism in society but nonetheless maintains its own ethnic gatekeeping arrangements, such as where psychological research on race disparities becomes dominated by White researchers who are then showcased as psychology's 'experts' in the field (McFarling, 2021). Psychology can reasonably be accused of 'virtue hoarding' with regard to racism (Liu, 2021), of trading in such 'symbolic capital' in preference to securing tangible improvements for society (al-Gharbi, 2024).

These inclinations toward virtue hoarding and elite capture are exemplified by psychology's engagement with the Open Science movement, itself a response to the field's so-called replication crisis (Hughes, 2018a). Efforts to improve the transparency and quality of psychological science are almost universally presumed to be meritorious by default. But this falls into a trap of equating 'openness' with righteousness. The problem is that Open Science is built upon the same platform as all other forms of science, and so retains the inequalities of academic colonialism (Bahlai et al., 2019). Marginalised scholars remain marginalised. Their research interests continue to be judged against WEIRD priorities, they still lack access to research budgets and institutional spending accounts, and they remain hopelessly priced out of so-called 'article processing charges'.

Indeed, marginalisation in psychology might in some ways be *worsened* by Open Science. An emphasis on the pre-registration of research trials caters well for studies with quantitative hypotheses that are grounded in extant literature. But it will yield much less benefit for qualitative or exploratory research, or for studies that focus on over-looked topics or populations (Grzanka & Cole, 2021). It is difficult to pre-register a study where the methodology has yet to be worked out, or to draw on an extant literature where no such literature exists. In a world where pre-registration is increasingly seen as an indicator of research robustness—and thus of a researcher's esteem—psychologists who specialise in exploratory or idiographic methods will quickly be left behind. And yet these are precisely the methods most likely to be used by psychologists whose research seeks to address the cultural, political, and contextual dimensions of social issues, and the psychological experiences of minoritised groups.

Further, the Open Science movement's ethos of unfettered peer-criticism might suit those who have social capital, but it likely creates risks for those without it. And all the talk about Open Science could be seen as just another form of conversational silencing, a way of *not* talking about who is privileged to do the important research in psychology, and why (Durrheim, 2024). In short, psychology's approach to Open Science is fraught with social justice implications even if it is ethically well-intentioned. Indeed, the very *fact* it is well-intentioned is part of the problem. Its virtuous signalling of forward-looking 'openness' distracts attention from how psychological science remains just as closed as ever to those it has traditionally excluded.

Finally, even the way psychologists discuss their field's WEIRD bias exhibits the dynamics of elite capture. What resembles a progressive discourse promoting diversity and inclusion simply persists in privileging the already privileged (Durrheim, 2024). As described at the outset, the 'WEIRD' construct itself involves conversational silencing—it omits to mention race, or Whiteness, as one of psychology's most conspicuous inclusion criteria. Moreover, its juxtaposition of *Western* alongside *Educated, Industrialised, Rich*, and *Democratic* problematically implies a semantic continuity across all of these notions. It suggests that to be 'Western' is something to aspire to—and, correspondingly, that to be 'non-Western' is akin to being uneducated, backward, poor, and oppressed.

But most troubling of all, the discourse on WEIRD psychology frames the entire issue within an internalist ideology. The discussion is centred on what WEIRD psychologists must do to fix things. Both problem and solution are envisaged from the perspective of those in dominant control, while the very fact that global privilege is concentrated within Western countries is essentially ignored. WEIRD psychologists define all the parameters: they get to decide who is WEIRD in the first place, what constitutes good science, and what to do about the challenge posed by their own over-representation. It is little more than a 'White saviour' narrative, in which psychologists from the world's richest (and White-majority) countries are portrayed as coming to the rescue.

37

Silence as Supremacy

Psychology's Whiteness problem involves a combination of straightforward ethnic and geopolitical imbalances alongside more subtle and conceptual biases. The more obvious issues include structural barriers that impede non-White participation in psychology, explicit hostility toward non-White contributors, and the distorted incentives of White-centred academia. The less obvious issues include psychology's undue emphasis on capitalism, its use of individualistic (rather than societal) explanatory frames, and its deeply embedded anachronistic ideas concerning the universality of American and European experiences. To a large degree all these issues are intertwined with one another. The colonialism that engendered Eurocentric theories of capitalist realism and a White standard for human psychology is the same colonialism that today ensures that White people, elite universities, and Western governments have disproportionate shares of the world's resources, as well as platforms from which to promote intellectualised rationales for their place within the status quo. Psychology, like everything else, operates with a pyramidal power-structure, one that just happens to correlate with society's most ingrained ethnic inequalities. Correlation does not represent causation, of course. Unless, that is, it does.

B. M. Hughes, *Psychology's Quiet Conservatism*,
https://doi.org/10.1007/978-3-032-07724-0_37

It is the nature of global relationships that the privileged are largely blind to their own good fortunes. With no little irony, they often consider themselves uniquely positioned to solve the world's problems. Consequently, while the distorted power structures of global psychology are frequently remarked upon by some of the most esteemed psychologists in White-majority countries, those same psychologists inevitably take it upon *themselves* to be the field's natural leaders in this time of crisis. Their (well-intended) initiatives to clean house frequently prove oppressive: centring White perspectives, paying lip service with deference politics, and focusing on technocratic tweaks rather than on sociopolitical upheaval—all serving to perpetuate the very problems that they wish to be seen (nay, *must* be seen) as addressing.

We have discussed throughout this book how a reliance on individuocentrism and reductionism push psychologists to focus on the personal rather than the political. In essence, these imperatives help ensure that psychology has nothing to say about the political at all. Psychologists are expected to explain the world and all its challenges in terms of individual human perceptions and interactions. They cling to neutrality, rationalising it as a form of robustness and believing it to be ethical—something to ensure that they judge data acontextually, for fear they might otherwise interpret things in their own favour. But this emphasis on scientific objectivity discourages the exploration of political or historical contexts. It produces a doggedly partisan epistemology (Teo, 2022).

Individuocentrism, reductionism, quantification, and all the other accoutrements of the scientific method are designed to maximise the neutrality of psychology's endeavours to transform evidence into knowledge. But the problem is that valorising objectivity so much implies it is in fact *possible* for a psychologist to put aside their own interests. It suggests that psychologists can be somehow immune to the partisan impulses to which 'ordinary' people (i.e., non-psychologists) so easily succumb. And because admissions of non-neutrality are usually regarded as red flags, signals that a study is 'unscientific', it also serves to discredit those researchers who are self-aware enough to explicitly acknowledge their own biases.

Scholarly neutrality when dealing with human affairs might be justifiable in an equitable world. But the arrangements of the world we live in today are very far from equitable. Whether or not White people find the phraseology comfortable, the world in fact is utterly 'White supremacist' in terms of how social, economic, and political power are distributed. White people reign supreme. Neutrality, therefore, is not really an option. A neutral stance in a White supremacist world is far from creditable. It can be viewed as ambivalence toward White supremacism itself, if not a form of White supremacism in its own right.

Psychologists often become energised when told that their field is popular. But seldom do they ask themselves *why* their field is popular. Why is psychology more heralded than sociology, anthropology, political science, or cultural studies? After all, these fields too seek to systematically study human behaviour in an effort to understand how people function in the world. Many will point to psychology's use of 'scientific' methods as the reason it stands apart from the other social sciences. But the effects of scientism on psychology's standing might not operate quite as they think.

Psychology's use of science might prove attractive not because of what it includes, but because of what it does *not* include. The other social sciences attempt to understand—and thus expose—the nature of our social systems. They explicitly set out to examine the roles and functions of institutions, political power, and societal injustice. Psychology, on the other hand, does the opposite. It actively *conceals* the nature of our social world from its calculations. Psychology is attractive to those with privilege because privilege itself lies beyond the boundaries of its remit. That Whiteness is part of that privilege is at the heart of psychology's identity.

All this puts into perspective the perversity of claims that psychology is liberally biased. According to its right-wing critics, what psychology needs to cure its ideological skew is, in fact, *more* representation for White conservatives. Their calls highlight psychology's core problems with Whiteness. White identities are so central to the norms of psychology that those who stand in favour of White privilege get to complain about how they are being treated.

Tackling racism in psychology is seldom led by White practitioners but typically left to those who suffer its consequences (Patel et al., 2019).

Racism is all too frequently portrayed as a burden carried by non-White communities, rather than something that tarnishes and demeans the whole of society. The individuocentrism of psychology chimes with those who, for self-preserving reasons, voice scepticism about the realities of structural racism, or display emotional ambivalence toward the experiences of people who live under racist subjugation (cf., Eddo-Lodge, 2017). This moral segregation positions victims as complainants, feeding into a stereotype of angry minority voices. It also signals an abdication of responsibility by White psychologists, who go about quietly availing of their majority privileges across North American and European society.

White silence is not a form of deference, it is part of the problem. It is itself a manifestation of White supremacy.

Part VII

Academic Exceptionalism and Psychology's Blind Eye

38

Internalising the War on 'Woke'

As the first quarter of our present century draws to a close, social conservatism is gaining ground around the world. Needles on political dials tilt ever rightward. Right-leaning nationalistic and authoritarian political movements are enjoying expanding electoral progress throughout Europe, the Americas, Asia, and Africa. With disaffected voters increasingly lulled by autocratic voices (Gessen, 2020), a confluence of populism, polarisation, and post-truth paranoia threatens to breach the flood-banks of democracy (Naím, 2022). With this right-wing resurgence has emerged an unfettering of hostility toward outgroups. Xenophobic opprobrium is no longer suppressed by circumspection. As the right ascends, political discourse has harshened. The unsayable has gradually become sayable.

Where once the norms of decorum would have discouraged the demeaning of difference or disability, it is now something of a trope for those on the right to mock, condemn, and articulate disdain for the very notion of diversity itself. Following the second Trump inauguration in 2025, with the new American regime declaring itself committed to eliminating all so-called 'Diversity, Equity, and Inclusion' (or DEI) initiatives from the federal government, agents were sent to sweep staff

B. M. Hughes, *Psychology's Quiet Conservatism*,
https://doi.org/10.1007/978-3-032-07724-0_38

offices of anything that might indicate a residual allegiance to wokism. Confiscated in one such raid on an intelligence agency headquarters was a supervisor's desk ornament bearing the slogan, *Be Kind to Everyone* (Merlan, 2025). No longer is it a social virtue to love thy neighbour or to show compassion to strangers. Instead the aim is to oppress any inclination you might once have had toward inclusivity, for fear that it might mark you out as unwilling to defend the in-group by whatever means necessary.

In this milieu, anyone who speaks up for the needy is now dismissed as some kind of supine snowflake, gullibly pandering to weakness. Those claiming rational insight on world events are delegitimised as purveyors of fake news, partisan politics, and skewed scholarship. Reasoned analysis of dominant group discourse is shut down using stock thought-terminating clichés—'Do your own research!', 'Don't be sheeple!'—as though the accumulated advice of peer-reviewed science and conventional news reporting cannot, on its face, ever be trusted. The witch-hunt language of anti-wokism declares that conventional sentiment in all its forms—news media, popular culture, and academic opinion—is skewed by a liberal bias. Social justice and scientific facts are jointly cast as tainted notions, the products of a corrupt elite of libs, who, for some never-quite-stated reason, are supposedly intent on destroying the tenets of society. What has long lumbered the field of psychology is now mirrored right across society, replete with all the customary rhetoric: 'We need a diversity of viewpoints', 'Liberal bias causes harm', 'Anyone who denies the existence of a corrupt and endemic liberal bias must be irrational, zealous, or deluded', and so on.

Liberal bias is essentially a euphemism for holding values that are supportive of liberal objectives, such as social justice and equality for all. In that light, where privileged people owe their status to financial *injustice* and hereditary *inequality*, it is pretty clear why many might view liberal bias as a threat.

Attacking systems of knowledge creation and transmission lies at the heart of right-wing ideology. Education and the culture wars have always been deeply intertwined. Universities are undoubtedly sites of privilege, but they are also places where the parameters of privilege are occasionally probed and critiqued. On their face, universities promote critical thinking and provide independent perspectives. They gather together thousands of people at a stage in their lives when they are most likely to be outraged by injustice (Müller, 2025). And despite all the paranoia that surrounds free speech on campus, universities—with their numerous debating clubs, student newspapers, and discussion societies—remain one of the few types of institution whose stakeholders enthusiastically participate in oratory, activism, and debate.

The conservative right's efforts to nullify campus insurgence has spanned generations, and yet, despite decades of defenestration and denigration, universities remain a thorn in conservatism's side. This is notwithstanding the fact, noted in Part I, that most modern universities operate as revenue-chasing neoliberal enterprises, are staffed by a gig-performing faculty who present their students with largely apolitical subject matter, and offer syllabi that are as couched in market-oriented value assumptions as any corporate strategic plan. Whenever resistance to society does emerge—where, say, academics and students rise up to challenge societal or global injustice—it is generally the universities themselves that push back the hardest, whether defending an employee accused of hate speech, or summoning the police, in full riot gear and with an armoured personnel carrier, to forcibly clear protesters (as infamously happened at Columbia University in 2024; Ramirez-Simon et al., 2024). Being anti-'liberal' is not merely a political emotion, it is a ruling principle necessary to keep the masses at each other's throats and away from the castle ramparts. University leaders therefore have an important role in augmenting conservatism. Privilege is as privilege does.

Academia is never immune to cultural trends. All scholarship reflects the wider culture from which it emerges. With the global rise of conservatism, anti-liberal academic sentiment has become more emboldened. Those psychologists who decry what they view as their field's liberal bias find their grievances amplified, elevated by a favourable shift in the Overton window. While continuing to cast themselves as surrounded by (hostile) liberals, their minority-status does not appear to limit their ability to attract attention.

Conservative academics can now often draw on active government support for their anti-woke crusades. The so-called liberal bias of fields such as psychology is under attack from the very top of society.

Governments in places such as Hungary, Argentina, Italy, and Poland have defunded—and in some cases shut down—academic departments, programmes, and funding agencies that are viewed as excessively liberal. In the US, federal authorities have taken extreme measures to obliterate 'liberal' research. In 2025, acting on foot of the newly elected president's anti-wokeness policies, the country's national science foundation vetted tens of thousands of its research grantees, aiming to identify studies that included words such as 'inclusion', 'discrimination', 'underserved', 'race', and even 'socioeconomic' in their paperwork, so that scientific investigation of 'red flag' ideas could be blocked (Mervis, 2025). Acting on the same instruction, the country's public health agency purged its website of several datasets concerning LGBTQ health (Wu, 2025), depriving researchers of access to years' worth of carefully collated statistics relating to youth risk behaviour and social vulnerability.

As explained throughout this book, psychology is constrained by many embedded features and practices that consistently steer its research toward conservative-friendly conclusions. These are now compounded by politically imposed logistical barriers and funding cuts. The impact of right-wing anti-liberal agitation has developed beyond merely shaping psychology's intellectual approaches. It now seeks to actively limit psychologists' practical capacity to pursue their science.

39

Good Science, Bad Science, Pseudoscience

Conservative calls for greater 'viewpoint diversity' suggest a belief that the mainstream is so intolerant of right-wing intrusion, it actively seeks to limit conservatives' freedom of expression (Duarte et al., 2015). However, many on the right are themselves engaged in restricting the freedoms of others. They are frequently reported as 'banning books, firing teachers who dare to mention that they're not heterosexual, outlawing women's health procedures, and making life a living hell for trans people' (Myers, 2023). In other words, conservative depictions of liberal intolerance are themselves frequently mirrored by corresponding forms of intolerance on the conservative side. The contradiction suggests either a capacity for casual hypocrisy, or a genuine blindness to double standards. Either way, believing oneself to be unassailably objective while reasoning through strong political convictions will undoubtedly create some level of cognitive dissonance.

An irony here is that according to many conservative academics, it is the left who engage in self-contradiction and denialism, whereas conservative argument, by contrast, more closely reflects the rationality and neutrality of scholarliness. It is even suggested that the Enlightenment itself was a product of traditional conservative thought. We saw in Part

II how Steven Pinker rebuked the political left for its anti-science leanings when dismissing liberal critiques of cultural colonialism (Pinker, 2018). His argument suggests that conservatives have the edge on intellectual prowess, whereas 'leftists' (to employ Pinker's term), due to their partisan impulses, are liable only to squander the hard-won fruits of scientific accomplishment. A prominent conservative commentator summarised the case on social media: 'Liberals have: feelings and emotion,' they wrote, whereas 'Conservatives have: logic, history, perspective, science, reason, facts, math, principles, ethics, values, dialogue & common sense' (Kirk, 2018). (Notably, the highest ranking reply to that post was from a commenter who responded: 'Yo, remind me again which side disagrees with [the] scientific consensus on evolution and climate change'; Christoph, 2018).

In relation to psychology, the conservative stance not only espouses its own intellectual merit, it also asserts that the wider field—by virtue of being so left-leaning—can no longer be regarded as a robust science. As Pinker puts it: 'leftist politics [have] distorted the study of human nature' (Pinker, 2018; p. 372). When faced with unwelcome research findings—for example, the recurring finding that gender-affirming care enhances trans people's mental health—conservatives need not reject the conclusions *on principle*, which would likely be seen as a partisan response. Instead, they can simply cast aspersions on the *quality* of the research, by complaining that liberal bias has degraded scholarly standards.

In this regard, they are aided by conspicuous evidence that mainstream psychology struggles with scientific quality (Hughes, 2016). Several methodological 'crises' have attracted attention in recent years, including problems caused by psychology's theoretical inconsistency, poor sampling, vague measurement, incomplete datasets, confirmation bias, misuse of statistics, under-reporting of findings, selective publication, and a traditionally loose approach to research standardisation (Hughes, 2018a). For more than a decade, efforts to replicate research published in the major journals of psychology have suggested that a majority of findings just cannot be reproduced (Open Science Collaboration, 2015). This should not be entirely surprising, as it has long been known that psychology studies tend to be weak by design.

Most have samples that are too small to address the hypotheses under scrutiny. Indeed, psychology's average 'statistical power' has been declining for decades, from a pretty poor 50% in the 1960s, to an altogether atrocious 25% in more recent times (Smaldino & McElreath, 2016). This suggests that many of the findings appearing on the pages of psychology journals are likely to be false positives. As alluded to earlier, one classic review of methodological flaws explained that, in fields characterised by small samples, irregular study designs, and a penchant for chasing 'hot' topics—in other words, fields *just like psychology*—we should, in fact, always expect a *majority* of published findings to be false (Ioannidis, 2005). In short, psychology research is far from unimpeachable. Therefore, just about any claim emanating from 'liberal' psychology can easily be challenged on the grounds that the underlying research was likely questionable, and probably parlous.

Of course, this latitude affords flexibility in all political directions. Just as liberal psychologists can interpret research in a liberal way, conservative psychologists can equally do so in a conservative way. For example, liberal psychologists can scrutinise the research on mental health outcomes associated with terminations of pregnancy and infer that abortions should be freely available (Major et al., 2009), whereas conservative researchers can examine the same literature and come to the opposite conclusion (Casey, 2010). Their only point of agreement will be that the studies are beset with methodological shortcomings (and, probably, that 'more research is needed'). Seldom, however, do authors explain that these shortcomings do more than just cloud the issue at hand. They create a *political* space where bias can contaminate the drawing of conclusions.

The point to remember is that neither side can lay claim to certainty. When it comes to the psychological science on detailed policy propositions, there is simply no such thing as watertight evidence. The inherent limitations of psychological research ensure that it is always possible to pick holes in whatever evidence is cited, because *all* studies are guaranteed to have fallibilities. Their lack of purity allows any critic to dismiss a position they dislike by declaring 'There is no solid evidence to support that!' (a cliché that is always literally correct if you consider evidence quality only in perfectionist terms, as an all-or-nothing idea).

This troubling technicality enables a form of double-speak that is commonly seen in contemporary politics, where commentators promote evidence-based approaches to some issues, but with others, dismiss as naive anyone who dares to suggest that policy should 'follow the science'. Statistics, data, and research findings are welcomed when it suits, but are derided when they prove inconvenient. The thinking employed here is not really 'based' on evidence, nor does it actually 'follow' any science. In fact the sequence is reversed. Evidence is selected *on the basis* of the individual's reasoning, not the other way around. And their decision on whether to rely on science *follows* their assessment of value-system implications, instead of preceding it.

If psychologists have garnered any insight from the study of decision-making, it is that people draw on empirical evidence only when it suits them. When it doesn't, they are happy to rely on principles, values, hunches, prior experience, authority figures, or common sense (Hughes, 2016). And, as is perhaps not emphasised enough, psychologists are people too. There is little reason (or evidence) to believe they are immune to the foibles of fast thinking that so hamper their fellow humans. Most psychologists acknowledge this human weakness, and point to it as a reason to remain vigilant in their work. But committing oneself to scientific objectivity involves a degree of overpromising (Clegg, 2022). Simply *trying* to be objective is insufficient to guarantee that subjectivity will successfully be overcome.

Assuming one's own political orientation to be inherently more objective than others is an especially treacherous delusion. It disavows the need for self-correction. The associated idea that science itself, with its rational objectivity, is somehow *synonymous* with conservative styles of thought—and, by extension, that 'leftist' reasoning is antithetical to good science—is especially pretentious. For one thing, it ignores the fact that the entire enterprise of scientific inquiry is premised on a willingness to change one's worldview should the evidence demand it, regardless of what authority figures might feel. This diverges significantly from the emphasis on preserving traditional understandings that social conservatism espouses. Conservatism, classically, is loyal to convention and sceptical of change.

In reality, the scientific method itself involves—indeed *requires*—a commitment to uncertainty. Confident claims that conservatives are rational—and therefore *right*—while liberals are erratic—and therefore *wrong*—overlook the tentative nature of knowledge. In reality both are right some of the time, and wrong the rest of the time. Logical scepticism, the belief that it is reasonable to question and investigate any statement purporting to represent factual information, is a fundamental assumption of science (Hughes, 2016). All claims to knowledge can, and should, be questioned, regardless of who makes them. The goal of science is not to eliminate doubt and bias, but to minimise them. Presumptions of certainty are nothing less than an active threat to science. Gauging a statement as likely to be true (or untrue) simply on the basis of the political views of the person making it is thoroughly unscientific.

The more interesting question, therefore, is not whether *science* has its own political essence, but whether *scepticism* has. Arguably, in promoting the questioning of established authority, scepticism is far more liberal than it is conservative. A sceptical orientation involves a willingness to challenge those in power and to countenance, and even embrace, radical change. In contrast, conservative values promote deference to established authority and portray those who call for wholesale change as reckless. This is as pertinent within psychology as it is across wider society: liberalism in psychology involves acknowledging the field's shortcomings and striving for improvement, whereas conservatism involves sticking with things as they are, on the basis that the old, tried-and-trusted ways are the best.

Conservatism purports to hold the higher ground on intellectual rigour and over-promises on objectivity. But these are forms of self-certainty that ignore the sceptical tenets of good science. If the field of psychology really does lack scientific robustness, then perhaps it is more likely as a result of institutional conservatism. Maybe it is a sign that psychology's practices are in fact *insufficiently* liberal—that is to say, insufficiently *sceptical*—rather than evidence that liberal thinking has somehow been allowed to run rampant.

40

Exceptionalism in Psychology

Thus far, we have seen how the field of psychology—while accused of liberal bias—is, in fact, structured so as to consistently reinforce a set of socially conservative outlooks. Its default approaches of individuocentrism, quantification, and reductionism combine to depoliticise, and thus disempower, the human experience. Its explanatory narratives are centred on the highly conservative notion of individual sovereignty, whereas critiques that implicate culture, society, or politics in the shaping of human welfare are marginalised. Basic theories of mainstream psychology embrace capitalist realism, pathologise political dissent, depict the natural world as subordinate to human needs, and compound negative stereotypes concerning out-groups. Psychology's focus on the primacy of individuals extends to its depiction of human suffering, which is seen as a person-level phenomenon, rather than as a social reality. It teaches us to expect, and thus to tolerate, other people's misfortune. In psychology's standard worldview, life is a personal struggle, and so those who enjoy social privilege need never feel responsible for their fellow human beings. None of this should be surprising given psychology's socially conservative history. The field

B. M. Hughes, *Psychology's Quiet Conservatism*,
https://doi.org/10.1007/978-3-032-07724-0_40

was born from theology, inspired by class conflict, and shaped by the eugenics movement. It evolved throughout a century of colonialism, and remains, to this day, beholden to the priorities of rich White-majority countries.

The inequalities of society are magnified in the profession, where they are compounded by the self-preserving impulses of privilege. Academic psychologists embrace all the neoliberal mechanics of academia—from grant accumulation, to university rankings, to journal impact factors, to scholarly citation counts—inflationary technocratic devices that systematically favour the haves in the mainstream at the expense of the have-nots at the margins. Meanwhile, applied psychologists adhere to training models that benefit candidates who enjoy pre-existing economic and cultural advantages. It is not for nothing that, in Europe and North America, psychologists, as a group, are disproportionately White and affluent. We are supposed to believe that psychology is a hotbed of runaway liberal bias, when in fact it is infused with a conservative dynamic that inhibits the achievement of social justice, by excluding from genuine consideration precisely those in society who would benefit most from it.

In some ways, therefore, psychology's many shortcomings might be seen as a good thing. Flaws should temper the enthusiasm with which psychologists promote their field. Every psychologist should be acutely aware that individuocentric studies conducted on WEIRD-focused samples and premised on capitalist realist assumptions are likely to be questionable. Therefore, surely every psychologist should speak about their field in the most tentative of terms, careful to caveat their claims with cautious references to the inherent limitations of their work. Surely they should exhibit what critical psychologist Thomas Teo (2022) refers to as 'epistemic modesty' (p. 8). But if anything, the opposite is more commonly the case. Despite the perennial state of 'crisis' that psychology has endured for over a hundred and twenty years—as early as the late 1890s, the academic journals were bemoaning '*die Krisis in der Psychologie*' (Willy, 1897)—its proponents are seldom shy when making grandiose claims on its behalf.

Extra, Extra

Much of this grandiosity arises from the field's inflationary publication practices and statistical approaches. A key problem relates to the fact that the average psychology study is underpowered. As mentioned above, statisticians estimate the average power of current studies to be around 25% (e.g., Smaldino & McElreath, 2016). What this means is that the typical sample size is adequate to detect only around one-in-four real-world effects. Consequently, a substantial portion of studies should be reporting null findings. However, in reality, nearly every published psychology paper reports a statistically significant result. This represents a problem of systematic exaggeration. It could be that most published results are actually false positives (as predicted by Ioannidis) or over-estimates of true effects. Or it could be that what appears in journals is a non-representative selection of what was submitted for publication, cherry-picked by editors to satisfy (commercial) publication demands. Whatever the exact reason, the result is all-round inflation. The never-ending parade of statistical significance creates a misleading impression of the actual state of the evidence (Hughes, 2018b).

Other publication patterns reveal different kinds of self-serving biases. For example, when a psychology study is conducted twice, its findings are far more likely to be reproduced if the replication is conducted by the *original* researchers than if it is conducted by *independent* researchers (Makel et al., 2012). Perhaps unsurprisingly, researchers appear to be less demanding when marking their own homework.

A similar pattern relates to what is known as 'therapeutic allegiance' bias. This is where studies of therapeutic interventions show larger effects when the researcher is themselves a practitioner of the type of therapy that is being investigated. For example, behavioural psychologists who conduct therapy research typically find behavioural psychotherapies superior to, say, cognitive psychotherapies. However, when cognitive therapists conduct such research, they usually find the opposite: in their studies, cognitive therapies will show larger effect sizes than behavioural ones. This type of bias is easy to spot when large numbers of studies are compared. One group of reviewers, having

examined over two hundred such randomised controlled trials, calcu-
lated that on average researchers appear to exaggerate the efficacy of their
preferred psychotherapies by more than 30% (Dragioti et al., 2015).
The strength of the effect led them to recommend that researchers
should in future be obliged to declare their therapeutic allegiances as
a non-financial conflict of interest.

And, of course, some exaggeration stems from rank insincerity on the
part of individual researchers. The manipulation (Craig et al., 2020) or
faking (Enserink, 2011) of data is not at all unheard of in psychology.
Much of the pressure to fiddle with findings comes from the neoliberally
competitive nature of campus science, in which gaming the system to
maximise metrics pays a far greater dividend than, say, changing the world
for the better (Szabo, 2025). Indeed, in the modern university, ambitions
to create a better world are often viewed as something of a distraction
from day-to-day business. Academic activism is seen as eccentric.
Management bean-counters are much more interested in bums on seats
and bank balances, and are left scratching their heads when somebody
mentions world peace. Such niceties, they feel, do little for the university's
bottom line.

Positive Self-Talk

But while some of the boosterism is structural, much is just old-
fashioned academic vainglory. Scholarly self-flattery has become
a modern norm. Back in 1974, only one-in-fifty journal abstracts
included complimentary self-descriptions in their text, for example,
statements describing the work as 'encouraging', 'promising', or
'novel'. But by 2014, this type of descriptor was appearing once in
every *six* abstracts, an increase in self-praise of nearly 900% (Vinkers,
Tijdink, & Otte, 2015). On its own the descriptor 'innovative' had
become 2500% more common, without any obvious indications that
the research being described was 25 times more groundbreaking.
Hyperbole has become part and parcel of contemporary academia—
indeed, of contemporary society—and psychology is no exception.

Psychologists join a wider modern culture that increasingly promotes extravagant self-projection (Alvesson, 2018), a profile-conscious society awash with elevator pitches, skills statements, and LinkedIn testimonials.

The rhetoric extends to the way psychologists defend their field from criticism. According to some, the fact that journals are stuffed with statistically significant findings is actually nothing to worry about. Rather, they argue, it shows just how *good* psychologists are at choosing which hypotheses to investigate (e.g., Gilbert et al., 2016). Instead of exposing suspicious patterns of confirmation bias, they suggest, the consistent hitting of targets merely reveals the sharpshooting acuity of psychology's research process. For defenders of psychology's honour, the true exaggeration lies elsewhere—it is all this talk of psychology being in crisis that is overblown!

This discourse of defensiveness can verge on the extreme. Some of those who wish psychology were simply left alone have decried the field's sceptical critics as 'shameless little bullies' (Bartlett, 2014), accused them of 'attacking science' itself (Edwards, 2017), and declared them guilty of 'methodological terrorism' (Letzter, 2016). One prominent professor referred to psychology's 'replication police' as the 'Stasi' (Gelman, 2019), while another writer juxtaposed methodological review with the horrors of the Holocaust, naming a chapter discussing critics of psychological therapies 'First They Came for the Communists' (Tuller, 2022). We might have hoped that psychologists whose work is criticised would welcome the opportunity of having their ideas tested in robust debate. Casting themselves as victims of bullying, terrorism, intimidation, and even genocide suggests, to put it mildly, a troubling resistance to scrutiny.

Exceptionalism as Nostalgia

Such defences are frequently framed within a narrative of linear improvement. Psychology's traditionalists portray its conventions as being calibrated perfectly for progress. It is for this reason that liberal bias is seen as such a scourge—it supposedly threatens to thwart psychology's inherent prowess, to dismiss the tried-and-tested practices that made the field so noble in the first place. Psychology ain't broke, the traditionalists feel, and so does not require fixing. To them, liberal bias is an interloping political ideology that impugns psychology's self-proclaimed objectivity.

Conservatives argue that before it was encumbered by liberal bias, psychology was just fine, because it 'worked'. It made 'progress'. But this just begs the question. Worked for *whom*? Made progress for *whom*? As we have seen throughout this book, the instrumental value of psychology will largely depend on who is wielding the instrument (Clegg, 2022). Most psychology research is subsidised by large institutions, such as grant-awarding agencies, research institutes, or universities—and so, by extension, by governments. Such bodies will always have their own aims and objectives, and will seldom choose to fund activities that undermine them. This is why research tends to meet institutional needs. Because of its reliance on material resource streams, the direction of psychology's output is shaped in ways that serve, rather than threaten, whoever holds society's purse strings.

For similar reasons, psychology's standard assessment methods and psychometric tests will typically align to the needs of society's power brokers. Aptitude tests confirm an individual's adequacy for schools and workplaces, locales that in the West have been shaped by the capitalist environment. Therapies and interventions are aimed at modifying people's behaviour so that it coheres with the surrounding (capitalist) world. In the end, we conventionally gauge psychology's capacity to 'work' or 'make progress' only insofar as its outputs contribute to the prevailing power structures of society.

Complaints that psychology has been ruined by the 'replication police' or 'leftists' amount to little more than an appeal that we turn back the clock—that we aim to make psychology great again. They are based on a conceit that psychology had it right all along, before all this talk of researcher bias, false findings, and methodological frailty—or indeed of individuocentrism, reductionism, capitalist realism, WEIRDness, and White privilege. The blind eye that so many psychologists turn to the field's shortcomings ultimately helps preserve the status quo, by encouraging deafness toward criticism and resistance to reform. In this regard, psychology's self-exceptionalism is more than just humblebragging. It is a delusion that feeds the field's inherently conservative tendencies.

Part VIII

Beyond 'Liberal Bias': Four Paths to a
Well-Adjusted Psychology

41

Rights and Responsibilities

Psychology's exaggeration of its own importance might seem more cringe than catastrophic. But for a field that purports to produce evidence-based knowledge, boastfulness is far from trivial. Saying things that are provably untrue is not a good look. Falsely claiming something to be factual simply because it places you in a positive light is the very epitome of intellectual corruption. It legitimises the faking of narratives regarding the state of the world. It cedes control of knowledge itself to the marketplace of reputation. It permits anyone to say anything they like, so long as it puts them at an advantage. Essentially, it admits defeat in the war on error. It renders truth optional.

Controlling the narrative has always been key to protecting the status quo. Elites could choose to suppress dissent by force, but that would create the risk of reprisal; far better instead to modify the information ecosystem in ways that promote compliance. This is why elites target education systems and media channels. With the right type of manipulation, both can be influenced to impart indoctrination, to '[make] system-preserving uses of information easy while rendering system-altering uses of information difficult' (Táíwò, 2022; p. 101).

Just as nineteenth-century British bureaucrats implemented education policies to reform colonised people into becoming 'English in tastes, in opinions, in morals, and in intellect' (Macaulay, 1835), so too are modern systems of schooling constructed to present the world with a type of psychology that serves its own propagandising purpose. The specious claim that psychology is intrinsically 'liberal' is intended only to make it less so. After all, the argument is that this already traditionalist field should bend even more toward conservative concerns. That such interventionism is called for tells its own story. The fear is that, left to its own devices, psychology as it ordinarily progresses would threaten to expose the true nature of the social world, and worse, inspire those who might dare to change it.

The claim of liberal bias in psychology is a conservative cry for help. It is a form of grievance politics. It is a complaint that psychology is no longer serving its intended purpose. The notion that 'equity' is unjust, that 'diversity' is exclusive, that 'inclusion' serves to ostracise, or that 'progressive' values are, in fact, retrograde, would amount to satire by any ordinary standard. But the extraordinary standards of modern conservatism involve compulsory contradiction. Truth, you see, is up for grabs. As we saw in Part I, even the claim that psychology is dominated by liberals is itself empirically questionable. The number of 'liberal' psychologists involved is likely far lower than asserted, and the most common examples of anti-conservative cancel culture turn out, upon closer examination, to be embellished or apocryphal. And as outlined throughout this book, regardless of the political views of individual psychologists, the overall body of work represented by modern psychology is, in fact, far from liberal. In other words, the claim of liberal bias is trumped up. It is another example of exaggeration in psychology, of propaganda in the service of self-preservation.

Even the distinction between 'conservative' and 'liberal' is largely contrived, an effort to polarise debate in order to leverage the murky human instinct for out-group opprobrium. The bipolar architecture of the implied political spectrum is arbitrary. 'If you are not with us, you are against us' has always been an unhelpful mantra, suggesting a bifurcation that is never really there. Even when used loosely, the dichotomy hides more than it reveals. The 'liberal left', for example, includes perspectives of the so-called far left, who reject capitalism as a concept and question whether democracy, as we see it today, really is the best system for society. There are very few out-and-out anti-capitalists, anarchists, or insurgent anti-globalists working in our universities, and yet we seldom see entities such as FIRE, or other critics of ideological bias in psychology, protesting their absence from academia. Calls for the imposition of 'viewpoint diversity' rarely involve proposals to foist Marxists on business schools.

In a similar way, the 'conservative' right comes in many shapes and sizes. It was a right-wing party in the UK that legislated to permit gay marriage. And there are plenty of white-collar, college-educated, credentialed professional psychologists, drawing above-average salaries, who can be relied upon to vote for lower taxes. The psychology profession, for the most part, engages in political quietism, apparently satisfied with society as it is. The group that appears to be missing from psychology is not 'conservatives' per se, but a particular subset of conservatism that takes its cue from evangelical Christianity (Ross, 2015).

And let us not forget that, when used in debates about psychology, the very nomenclature of 'conservative' and 'liberal' is seldom that philosophically sophisticated. Rather, it is a WEIRD-centred social construction, a dog-whistle tuned for constituents who live in North America, the UK, and Europe. From a global perspective, political orientations frequently revolve around issues unrelated to the Western left–right divide. Beyond the WEIRD bubble, political discourse spans many varied ideologies, such as distributism, agrarianism, separatism, Jihadism, reformism, Pan-Africanism, collective capitalism, and cosmopolitanism, to name just a few examples (Seibt et al., 2015). The localist tenor of the liberal–conservative dichotomy, and thus of the claim of liberal bias in psychology, is evidence of its partisan purpose.

The intention of the complaint is not to inform, but to inculcate, to impose a particular narrative. The culture wars are in full swing and attacks on psychology's supposed liberalism are collateral damage. The critics who want psychology to be less visibly 'liberal' share common cause with those who want the Disney corporation to have fewer gay characters in its movies. It is a paranoia about symbols rather than about political philosophies. One side needs to 'win' over the other, and whatever it takes to conquer disputed territory is fair game. Thus the dominant couch their complaints in emancipatory terms, adopting the discourse of viewpoint epistemology to self-identify as victims of persecution. They say they want greater viewpoint diversity in psychology while mocking the very word 'diversity' when used in just about any other context, effectively trolling the libs by throwing their own language back at them.

And so in psychology, as in every other part of modern culture, control of the narrative is core to the enterprise, perhaps even more important than generating knowledge per se. As we have seen, psychology is less liberal than might be supposed. It is firmly ensconced in a conservative disinformation bubble, a White person's science that addresses First World problems by drawing on traditionalist values. The entire field could be said to be an example of conversational silencing, that tactic of talking about Topic A ('personal responsibility' and 'liberal bias') as a way of not talking about Topic B ('societal inequality' and 'cultural revolution'). In short, the instrument of psychology may indeed be wielded by well-meaning intellectuals and practitioners, but its effect is to preserve a very narrow form of privilege. It is, in its current dynamic, a tool of the status quo.

Psychologists can take control of this tool if they want to. It is certainly their duty to try. Noam Chomsky—the fabled American philosopher, linguist, activist, political theorist, and psychologist—defined the responsibility of intellectuals as 'to speak the truth and to expose lies' (Chomsky, 1967). For Chomsky, such responsibility is a function of privilege. Or, to avoid language that conservatives might find triggering, the responsibility is owed *in return for the many benefits that intellectuals receive from society*. Psychologists, for example, not only possess social (and often financial) capital, they have been trained, at great public expense, to be proficient researchers and communicators. They have a duty to provide a return on society's investment. They owe service to those who have subsidised them.

To close, let us consider four paths toward fulfilling the responsibilities that come with the privilege of being a psychologist. Four paths by which we might enhance the discipline of psychology, address the field's *actual* biases, and, by extension, unleash its power in the service of democracy, integrity, and justice. Intellectuals are professional purveyors of truth. Chomsky explains that when truth is under attack, it is always far easier to stay quiet than to fight back. This is the very reason we must not do so.

42

Path #1: Effortful Diversity

As mentioned, hostility toward diversity is part and parcel of the conservative critique. Not only do conservatives voice their objections to the implementation of diversity-promoting policies and programmes, many adopt a zero-tolerance approach to even the mention of the idea. Recall, from Part I, how critics of psychology's purported liberal bias have ridiculed such terms as 'inclusiveness', 'community', 'intersectionality', and 'safe space' for allegedly lacking scientific robustness (O'Donohue, 2023). And compare that with the previously mentioned US national science foundation's lexical review of research projects, which sought to cull any that employed 'red flag' phrases identified by the government as indicating a now suddenly outlawed interest in diversity (Mervis, 2025).

In reality, of course, it is not diversity per se that is the problem. What is in play are the politics of in-group supremacy, whose motivating fear is of an unravelling of inequality that might lead to increased competition and a more meritorious distribution of resources. In that nightmare scenario, those who benefit from current inequities are destined, on average, to lose out. Hence their inclination to demonise any discourse that even references the possibility.

B. M. Hughes, *Psychology's Quiet Conservatism*,
https://doi.org/10.1007/978-3-032-07724-0_42

In psychology, the argument is not just that diversity is a vague concept, but also that it is a political one. The subtext here is that politics and science should not mix. Psychologists are especially prone to being persuaded by this because of their field's intrinsically depoliticised orientation, in which all problems are supposed to be viewed individuocentrically, with political and societal factors left ignored. Psychologists are trained to believe that in their research programmes, but also in their professional lives, they should leave politics to the politicians (Degerman, 2019b). The problem with this is twofold.

First of all, political forces affect people's lives to such a degree that any study of the human experience that *lacks* a political dimension will inevitably be incomplete. As discussed in Part II, a psychology that omits political considerations will produce highly misleading accounts of why people end up happy or sad, motivated or apathetic, thriving or at risk, and so on. For example, it will overlook the fact that the leading cause of major depression is not negative self-talk or low serotonin, but economic poverty.

The second part of the problem is that no matter how objective they try to be, all psychologists will have their own range of political preferences. It is pointless to pretend to be apolitical. Transparency demands not the pretence of neutrality, but that each psychologist acknowledge their positionality on any issue that comes before them on which they hold a view. Subjectivity is not a professional sin. And believing in a better world is not a faux pas. As such, if a psychologist wishes to stand up for social justice, equality, or diversity, it is by no means incompatible with their professional orientation. Psychologists are quite entitled to support such causes. Indeed, it would be insincere of them—and thus *un*professional—to act as if they do not.

Despite all the blather surrounding the term in contemporary commentary, the fact remains that 'diversity' refers to a simple principle, whereby we try to ensure that all people are represented in what is happening. The *absence* of diversity involves neglect, disempowerment, and passive marginalisation. The *opposite* of diversity is exclusion, monoculture, and human homogeneity. It is difficult to see how any of these outcomes could be argued to be morally desirable. And yet psychology, as a field, struggles with diversity in both theory and practice: its curriculum is constrained by WEIRDness, while its profession is disproportionately White, Euro-American, and from affluent backgrounds (not to mention cisgender, non-'disabled', and middle-aged). Correcting psychology's lack of diversity is not just a matter of personal representation, it is necessary for good science. It will be impossible to produce empirical insights into the nature of human experiences so long as psychology's sampling of such experiences is skewed. Individuocentrism and depoliticisation narrow psychology's explanatory reach. Good science demands *more* diversity, not less.

Beyond that, notions such as equality and justice are empirically quantifiable. Equality refers to whether one person has the same as another. Justice refers to whether something that applies to one person in one situation applies consistently to other people in the same situation. In methodological language we would refer to these concepts as matters of external validity. Specifically, we would ask, are the core indices stable across cases? It is entirely reasonable for psychologists, as scientists, to examine these issues, and to do so using empirical methods. And given that human welfare is itself a core part of psychology's subject matter, it follows that issues relating to diversity are inherently important for psychologists to discuss. A failure to discuss them would detract significantly from psychology, rendering it a weak(er) science.

Some psychologists might worry that adopting a vocal position on matters of diversity will simply attract more flak from the field's conservative critics. This is probably true. But conservatives already label psychology as 'liberally biased' anyway and are unlikely to stop doing so any time soon. It is perhaps time to reclaim the point on psychology's behalf. It is time to promote diversity with effort. This is because what conservatives think is liberal bias is not a bias per se. It is a logically reasoned and empirically informed stance that effortfully emphasises the diversity of human beings. It is essential for a proper understanding of our subject matter.

On a practical level, the effortful promotion of diversity will require meaningful engagement by psychologists across their roles, beyond any contributions they might make as guest speakers to their HR departments' DEI seminars. Effortful diversity should become a professional orientation. It should strive for humility, encourage intergroup dialogue, and promote a panoply of voices, rather than the cosplaying of social justice by White-saviour professionals (Malik, 2024). It should be active—even activist—rather than passive. Diversity does not happen spontaneously, nor does it champion itself. Psychologists, especially those from majoritised groups, should seek to ensure that all activities they are involved in reflect the principles of diversity and inclusion. They should point out to others, repeatedly if necessary, that committees, panels, teams, and representative groups should seek to mirror the constituencies they are drawn from and the communities that they serve.

When promoting effortful diversity, psychologists should acknowledge that activism itself is stressful. Choosing to internalise the burdens of others in the face of apathy and cynicism requires commitment and persistence. It may create feelings of despair, given that not all problems can be solved by an individual. It can present tangible risks to psychologists whose own employment is precarious, or who lack the social capital that is sometimes needed to take a public stand. That said, where a psychologist can exercise their political voice, they should endeavour to do so loudly. It will elevate their sense of purpose, and can itself be seen as a type of self-care (Nadal, 2017).

Effortful diversity will ultimately require some degree of self-sacrifice. It makes little sense to advocate for a more equitable distribution of privilege without a willingness to give away (at least) part of one's own. This might sometimes mean stepping away from prestigious positions so that others who might otherwise be unfairly excluded can step into them. In fact, given the lack of diversity across the field of psychology, it will most likely mean regularly having to do precisely that. But self-sacrifice need not be seen as an aversive prospect. It is not synonymous with self-harm. Self-sacrifice is an investment toward bringing injustice to heel, because an injustice for one is injustice for all. The quid pro quo is the opposite of what conservatism preaches. With diversity, everyone wins. Without it, everyone loses.

43

Path #2: Constructive Action in Education and Academia

Many psychologists pursue their efforts in education and academic settings. There are two strands to this. Firstly, there is the organisational or logistical strand, where a psychologist might consider their behaviour and actions while at work. And secondly, there is the small matter of the coalface—where psychologists, through their research and scholarship, carve out new contents for their field, and deposit them piece by piece onto the conveyor belt of human knowledge. This second strand, where epistemic gravity drags psychology's depiction of human beings toward persistently conservative worldviews, requires particular attention and we will return to it below.

As mentioned repeatedly so far, many academic psychologists already engage in the promotion of liberal values. Of course many also do not, and it is generally the case that the majority are politically quiet. For the reasons outlined above, psychologists might wish to consider whether indeed they have a duty to generate some noise. After all, not only do they make a living from talking about human welfare (and so would surely wish to cover the topic comprehensively, rather than in a reductionist, and thus impoverished, manner), they are also in the truth-generation trade, and it is patently obvious that truth does not

© The Author(s), under exclusive license to Springer Nature Switzerland AG 2025   **309**
B. M. Hughes, *Psychology's Quiet Conservatism*,
https://doi.org/10.1007/978-3-032-07724-0_43

come forth easily in this modern world. Indeed, submerged as they are in misinformation, disinformation, echo chambers, and fake news, many modern audiences seem resistant to shared truths. So even at this superficial level, psychologists have a duty to stand up for their own processes. They have a vested interest in promoting empirical literacy among those who might seek to learn from them.

As trite as it might sound, academic psychologists must therefore lead by example in the classroom. They should explain to learners the nature and impact of individuocentrism, reductionism, and quantification. They should raise the concept of capitalist realism. They should explore the possibility that psychology might have accumulated views on social issues that are fundamentally incomplete, and that these views reflect the preferences and interests of those in the population who have had the privilege of shaping their field. They should call out psychology's WEIRD-centrism and explain its colonialist roots. And they should own up to the fact that much of what happens in psychology today reflects societal inequalities and prejudices. This will involve directly engaging with hot button topics, including race, gender identity, homophobia, classism, sexism, ableism, and politics, subjects that are rarely ever covered in contemporary psychology curricula, except for the occasional special-topic elective class that the majority of students avoid.

How material is covered is as important as what is covered. Academic psychologists may feel ill-equipped to discuss controversial issues with their students. The solution is not to avoid these topics, but to upskill in the pertinent pedagogies. Professional development in academia necessitates keeping abreast of teaching trends. As psychology students become hungrier for their teachers to address the burning issues of the day, and as psychologists become more aware of their field's controversies and constraints, the ability to manage challenging classroom discussions should no longer be regarded as a luxury. It should be seen as an essential skill.

This applies not only to psychologists whose specialisms deal with traditional 'social' issues. Even those teaching ostensibly apolitical areas will need to get on board. Neuroscientists should be ready to explain why the vast majority of brain images are of White people's brains (Laird, 2021). Cognitive psychologists should be able to discuss how research on topics such as attention was spawned by the military–industrial complex (Holmes, 2020). Biological psychologists should encourage reflection on how reductionist theorising accentuates victim-blaming stereotypes. Research methodologists should consider the dangers of presuming quantification to be objective, as well as the eugenicist roots of their psychometric and statistical techniques.

Those who teach and write about psychology should balance the visibility of both mainstream and critical perspectives. Rational scepticism, of the kind ordinarily seen as essential to science, should be embraced rather than shied away from. Students should see that sceptical inquiry is healthy rather than harmful, and that the questioning of assertions is wholesome and not 'toxic'. The clinical orientation of accepting whatever clients say, on its own terms and without challenge, should not be extended to whatever is said in journal articles or to the interpretation of claims to fact. Peer-review should be positioned in terms of quality control and not professional decorum. Indeed, issues such as tone policing (or calls for viewpoint diversity) should be recognised as manifestations of social power and bias, as matters as pertinent to empirical rigour as they are to social justice.

Most academics already consider themselves to be committed to critical thinking. But psychologists should be particularly cognisant of the pitfalls of self-delusion. The third-person effect—the belief that logical fallacies are things that affect *other people* more than the in-group—is a particular threat, especially given psychology's tendency to inflate its own record of dealing with its various 'crises'. Psychologists can be fond of recommending their own practices to others, as if all the problems have been eliminated from their own research. But it is an exaggeration for psychologists to claim to have dealt with the replication crisis simply because they have talked about it at conferences. Critical thinking is little more than an academic cliché unless *self*-criticism is prioritised.

That said, hasty dismissal of non-conforming opinion itself lacks critical acuity. Academics and educators should always promote an 'equality of intellectual authority' (Longino, 2002), where speakers have a right to be heard regardless of their status (Clegg, 2022). This is especially important for teachers, who otherwise might find that students internalise a sense of their own 'natural' classroom inferiority. The principle extends to research too. Psychologists should be receptive to the idea of participatory research practices, such as qualitative methods that cultivate voices and narratives. But even researchers who conduct highly quantitative studies on behaviours, cognitions, or experimentally derived indices should afford participants the right to influence research directions. So-called 'patient and public involvement', or PPI, is an important principle for progressive scholarship (Rose & Beresford, 2024). Its relative rareness in psychology contributes to the field's conservative instincts, which see epistemic decision-making as centred within educated elites.

Academic work involves a high level of engagement with bureaucracy. This domain is no less a site for constructive action than the classroom or laboratory. Academic psychologists will have ample opportunities to improve the world by conscientiously contributing to the workings of university panels, committees, and governance processes. Such bodies require vigilance because institutional checks and balances frequently fail to account for cultural issues, social power, and exclusion. Those who smash through academic glass ceilings are often heralded by their universities as pioneers, without those universities ever doing much to deconstruct their discriminatory architecture. Psychologists, with their (supposed) expertise in perception and reasoning, are well placed to advocate for meaningful change. To do this will require a willingness to part with conditions as they are, and resistance may well be encountered. As pointed out by the feminist psychologist Naomi Weisstein, the academic status quo is 'a social conspiracy against the powerless' (Weisstein, 1993). Collective force is required to disrupt it.

Perhaps a more orthodox exercise of bureaucratic dexterity would be to enhance the psychology curriculum. One option would be to establish dedicated courses, suitable for undergraduates, that teach activism and social change (Nadal, 2017). Classes that highlight the interplay across psychology, history, and politics, or that deal with the impact of value systems on psychological research, can help encourage useful perspectives on psychology's context, while building students' confidence to become conscientious consumers and practitioners of their field. Developing a standard whereby such tuition is seen as core to an education in psychology helps to ensure that future generations of psychologists will have at least thought about these issues. It also normalises the practice of self-critical inquiry regarding the field's political and cultural dimensions.

44

Path #3: Constructive Action in the Public Square

It is sometimes said that academic activism flounders upon contact with real people. The world outside the ivory tower cares little for postmodernism, epistemology, or hermeneutics. In fact, in the public square, there is often a strong pushback against the implied superiority of academic knowledge, especially from the conservative right. Long before the Brexit campaign soundbite suggesting that British people had 'had enough of experts' (Mance, 2016), an American senator dismissed criticism of his poor understanding of the world by gleefully declaring, 'I'm not a scientist, man' (Miller, 2012). (That same senator would later be promoted to become the nation's secretary of state.) Conservative animosity toward universities often correlates with a wider mockery of intellectualism.

Intellectualism is not a body of knowledge. It is a stance about the power of human intellects to work out what is true (Broudy, 1954). The fear that such a thing might pose a threat, especially to elites, has been prevalent throughout much of history. Anti-intellectualism, as such, comes in many shapes and sizes. In the 1990s, sociologist Daniel Rigney (1991) described three main types of anti-intellectual attitudes, drawing on some earlier ideas by Richard Hofstadter, an American historian.

© The Author(s), under exclusive license to Springer Nature Switzerland AG 2025   **315**
B. M. Hughes, *Psychology's Quiet Conservatism*,
https://doi.org/10.1007/978-3-032-07724-0_44

Rigney called the first type 'religious anti-rationalism', which refers to the belief that emotion, or sentiment, is always preferable to reason. This is the view that rational critiques of authority threaten to undermine traditional society and to collapse the moral order of the world. Rigney also identified a form of anti-intellectualism that he called 'unreflective instrumentalism', the idea that knowledge is important only if it generates a material profit. In this view, any form of thought that does not yield a significant and immediate payoff is seen as lacking value. Rigney called his final type of anti-intellectualism 'populist anti-elitism'. This view is usually promoted by those with positions of privilege. It states, on their behalf, that the 'ordinary' people really have had enough of experts.

In modern times, we can perhaps suggest some additions to Rigney's list. For example, today we also frequently encounter a type of anti-intellectualism that we might refer to as 'childish anti-nerdism', where playground insults become lifelong stereotypes, and where the scorning of nerds, geeks, dorks, high-achievers, know-it-alls, and gifted peers becomes normalised in adult life (e.g., Siegel, 2025). We also often see an attitude that we could call 'above-it-all anti-wokism', in which educated liberals feign disregard for popular progressive discourse as a kind of flex, not wishing to align themselves with what others ridicule as po-faced self-righteousness (e.g., al-Gharbi, 2024). In this form, anti-intellectualism is used *by* some intellectuals to discourage *other* intellectuals from making such a fuss. And we can also identify a form of 'anti-intellectualist scepticism', a critical stance that decries how intellectualised discourse is frequently hijacked by elites to camouflage bigotry toward minorities (e.g., Kendi, 2023).

Intellectualism is often disparaged as a way of shooting the messenger, for fear that the message might otherwise be received. Attacking intellectuals is a useful way to devalue the assessments of 'experts' regarding Brexit, the concerns of climate scientists regarding ecological demise, or the advice of public health officials in relation to lockdowns, masking, and vaccines. When audiences are primed to reject intellectualised discourse, anyone who tries to explain how societal structures perpetuate race-based inequities can easily be vilified by blanket rejections of 'critical race theory'. Overall, flooding the zone with cynicism towards intellectuals makes it harder for intellectuals to win arguments using evidence or logic. It drowns out rationality with rhetoric. This is why psychologists' advices on such issues as abortion, mental health, smart-phone use, teenage sexuality, parenting, gender-affirming care, or micro-aggressions are all frequently met with a cacophony of anti-intellectual slurs, denigrating academics for their cluelessness, their poor morals, and of course, their liberal bias.

Chomsky's imperative that intellectuals have a responsibility 'to speak the truth and to expose lies' emphasises ideas rather than form. Intellectuals do not need to use long words or convoluted narratives to state their case. Indeed, truth is more impactful when it is expressed in simple ways. Simpler language is more likely to be understood by a larger number of people. And simplicity is like kryptonite against anti-intellectualism. Critics tend to flounder when faced with straightforward arguments. They much prefer to have some obfuscatory verbiage to ridicule.

For Chomsky's dictum to have its desired effect, intellectuals, includ-ing psychologists, must communicate with society as a whole. Publishing academic journal articles about critical theory—or, for that matter, verbose books on political bias in psychology—can only ever be part of the endeavour. Therefore, to reach the widest audiences, psy-chologists should be prepared to speak in several tongues. Most psychol-ogists find communicating outside the academic domain particularly difficult. Not all their struggles relate to anti-intellectual pushback; some arise from woefully inadequate communication skills, honed to opacity over years of scholarly immersion. So an initial requirement for con-structive action in the public square is that psychologists upgrade their communication. This is, of course, easier said than done. But improve-ment is never impossible, and opportunities to learn are plentiful.

The teaching that happens in public settings outside traditional school-based systems is sometimes referred to as 'informal' or 'free-choice' education (Falk & Dierking, 2002). It is quite different from classroom chalk-and-talk and requires its own particular skills. It spans the gamut of what used to be called 'sci comm', or science communica-tion, and incorporates such outlets as media interviews, public talks, magazine articles, blogs, and social media posts. Psychologists will be better able to reach wide audiences if they reflect on and understand how these media work. Emotionally intelligent interpersonal skills will also be important, such as the ability to see things from other people's perspectives. Persistence, too, will go a long way.

But perhaps the most important orientation required for success is to understand that communication in the public square employs a dialect quite unlike the one used in academia. Communication with non-specialists requires not that psychologists *simplify* their language, but that they *translate* it into the public patois, contextualising their narratives not with citations of 'study X' or 'famous-thinker Y', but with reference to common cultural touchpoints. This need not involve cheesy allusions to memes or pop culture. Rather, psychologists can connect with their audience by invoking shared aspects of people's lives, such as their health, their families, or their cultural history. People are usually most interested in how new ideas relate to their own experiences. As such, when planning how to communicate outside of academia, the audience's own experiences should be the starting point.

Of course, not all audiences will be receptive to what psychologists have to say. A major challenge with public communication is learning to navigate polarisation without succumbing to centrism. Or, in other words, to face down the heat without selling out on your convictions. Most communication specialists will advise speakers not to be provoked by audience anger; the essential skill, they say, is to *engage* rather than to (further) *enrage*. Nonetheless, it is always going to be difficult to change fixed minds.

But despite appearances, most people do not actually occupy polarised positions. Their hearts, and their ideas, are rarely that hardcore. This passive majority might well be inclined to keep their heads down politically, but they will be far more willing to listen to new ideas than their vocal counterparts at the extremes. All told, psychologists hoping to speak the truth and expose lies should seek to conquer this quiet middle ground. Should the passive majority recognise that their quietism is counterproductive, this alone might prompt them to become vocal (Nadal, 2017). And according to empirical studies, the silent majority are there for the taking. A large group can be won over by convincing just a subset (Monbiot & Hutchison, 2024). Research puts the social tipping point at around 25%—once a quarter of people commit to change, the rest quickly follow what they see as the emerging trend (Winkelmann et al., 2022). So psychologists should not be discouraged by pushback per se. To attempt to persuade everyone is an impossible ideal, but it is also unnecessary.

Psychology research is of most value when it is promulgated widely and where academic content is applied to real-life needs, such as campaigns or lawsuits (Fine, 2016). As such, constructive action in the public square will require not just an accessible communication style, but also physical access to ideas. With the ongoing fragmentation of traditional (or 'legacy') media and their monopolisation by vested interests, access to publication is ever harder to guarantee. For this reason, an increasing number of writers now choose to control the availability of their own ideas using complementary publication platforms, such as personal blogs and newsletters. Blogs can also serve as repositories for posting academic papers where pre-print licences allow it (such papers can also be posted on several large-scale repositories especially designed for this purpose).

How information flows is essential to whether the status quo is challenged or preserved. This is why oligarchs are in the business of buying up mainstream and social media to use as their personal propaganda platforms. Public communication is necessary for psychologists to combat the anti-intellectual myth that their field is liberally biased, so that they can debunk false claims about what psychology stands for and unleash its true progressive potential.

45

Path #4: De-privileging Psychology

The claim that psychology is liberally biased is rhetorical misdirection. Stereotyping the field as ultra-liberal distracts us from ever thinking about it as a force for conservatism. Few psychologists ever stop to ask why exactly their field prospers so well within the Euro-American, White-majority, capitalist echo chamber. The answer is because it is highly compatible with the status quo. Psychology speaks largely from the perspective of privilege. Its methods and theories may permit some tweaking around the edges, but mainly they serve to keep society the way it is. Tarring the field with the brush of 'liberal bias' is useful conservative propaganda. It depicts liberal psychologists as zealous, exploitative, and irrational bullies who deserve to be knocked from their perches. It encourages middle-ground psychologists to believe that their rightful role is to quietly disconnect from politics. And it allows conservative psychologists to maintain psychology's capitalist and traditionalist worldviews, emboldened by their putative vindication in the culture wars. This is how conservatism in psychology ultimately prevails. It convinces the world that psychology's problem is not structural privilege and racism—but *people who complain* about structural privilege and racism.

B. M. Hughes, *Psychology's Quiet Conservatism*,
https://doi.org/10.1007/978-3-032-07724-0_45

A key agenda for improving matters is to challenge the cliché of 'liberally biased' psychology, a stereotype that many psychologists themselves buy into. One way to approach this matter is to see it as an empirical proposition. Just like any data-based claim, psychologists should be able to critique the quality of the evidence. And as we saw in Part I, the evidence is extremely weak. The numbers don't check out. They are derived from a tiny set of cherry-picking surveys and quasi-anecdotal reviews. Most actual psychologists are in fact *apolitical*, if not politically-allergic. In addition, the concept of liberal bias in this context is poorly defined. It emanates from a neoliberal academic tradition. It conflates traditionalism with normality and problematises dissent as 'bias'. And it is based on a false dichotomy drawn from the media discourse of the world's richest Western nation. The liberal bias claim is not just empirically weak, it embodies much of what makes psychology so conservative in the first place. It is highly reductionist, premised on capitalist realism, and utterly WEIRD-centric.

Science and Scientism

As pointed out in Part I, a conspicuous irony arises from the 'liberal bias' claim. It is a core part of the complaint about liberal bias that a lack of political diversity is said to damage the quality of science in the field (Duarte et al., 2015). Critics assert that psychology's lack of conservative voices undermines the validity of its research. And yet, the 'science' used to demonstrate the existence of this dreaded bias is itself extremely low-grade. This alone should be emphasised much more frequently, given how the liberal bias claim is repeated far and wide despite its poor grounding in evidence. The fact the claim has any currency at all owes more to confirmation bias than to corroboration. The sheer flimsiness of the case for the prosecution has been in plain view, in the public domain, for decades.

Many psychologists are frankly embarrassed at the fact that others accuse them of 'liberal bias' (or worse, of being 'woke'). They feel demeaned that the integrity of their scientific work is being called into question. But this reasoning misses many points. Firstly, it is oblivious to the various ways in which psychology is, in fact, far more conservative than liberal. And secondly, it fundamentally misunderstands the nature of public perception and of attitudes towards their profession. Psychologists are not deferred to because of their impressive data. They are deferred to because they are recognised as having 'symbolic capital' (Bourdieu, 1993). Their status as public authorities is derived not from the quality of their empirical evidence, but from their credentials, titles, institutional affiliations, and scholarly name-dropping. Wokeness serves as a further form of symbolic capital by enabling proponents to project themselves as being on the right side of cultural history (al-Gharbi, 2024). The impact of symbolic capital reflects the expediency of public opinion. It provides psychologists with a shortcut to respect, a gravitas that implies rather than demonstrates their authority.

As we have seen, psychology's tools actually maintain the master's conservative house. It is the very fact that psychologists are recognised for their symbolic capital—for their rhetoric more than for their research—that opens their field to accusations of liberal bias. How people talk is often more impactful than what they say. When psychologists couch their conclusions in progressive jargon, they can easily be accused of being out of touch with 'ordinary' people. A common criticism of liberal causes such as identity politics (apart from the aspersion that special-interest groups who advocate for their own rights are somehow diminishing the rights of others), is that genuinely vulnerable people neither think nor speak about their lives in woke terms. The fact that marginalised groups do not use the same language as academics is taken as evidence that the academics are out of touch. But differences in vocabulary are cosmetic. The suggestion that 'ordinary' people are uninterested in ideas of social justice—that they are unconcerned by systematic exclusions, victim-blaming narratives, or structural racism—is far from convincing (Sarkar, 2025). If psychologists avoid raising these matters for fear of further ridicule, then it is 'ordinary' people who will stand to lose the most. But that, of course, is the point of the ridicule.

To truly de-privilege psychology, we need more than a commitment to better politics. We need to fundamentally change how psychology frames its work. As mentioned, while being a somewhat diverse science, psychology is dominated by a particular mainstream. Its ongoing focus on individuocentrism, reductionism, and mathematical reasoning is highly problematic. The way psychologists depict the world, using population sampling, variables, and statistical significance, is an arbitrary approach, one that originated with the eugenics movement. By and large, psychologists are acculturated to construct their research questions using an architecture of statistical procedures that was designed for the sinister agenda of separating people into groups, based on race and social class, for the purposes of rank-ordering them. The chore of representing concepts numerically and then making inferences based on mathematical analyses of those representations has become an unreflective practice (Tafreshi et al., 2016). We unthinkingly employ the eugenicists' toolkit to 'do' psychology, and we consider it just the ordinary way of things.

Ring-fencing human groups and representing their traits as mean averages, before formulating hypotheses on that basis, is not a natural way to frame questions about human experiences. It represents a fundamentally restrictive paradigm. Perhaps it is no wonder that the corpus of psychological knowledge is structured as if people exist as individuals within tribes, afflicted by computable group-level deficits and competing for status within a species-wide hierarchy. Perhaps it is no wonder that the outputs of psychology directly permit, and propel, tribal politics.

Quantitative research is hugely important to the field. But it is not the only way to expand psychological knowledge. To de-privilege psychology we should broaden our understanding of the research process. The scientific method ensures rigour for certain types of hypothesis, but there are other ways to ask and answer questions about human beings. The robustness of statistical hypothesis-testing does not, in itself, trump other forms of knowledge. The rhetoric of scientific certainty is a social performance that leads to the narrowing of disciplinary conventions (Koch, 1981). It feeds academic protectionism.

Academic psychology operates within internalist hierarchies, in which certain research practices and types of evidence are privileged over others. It also exists within a university system in which institutions compete for rankings linked to (for example) the impact factors of the journals in which their researchers publish. All these forces drive a homogeneity of practice that is shaped by the needs of those who are privileged enough to determine the rules of engagement. The conveyer-belt that provides psychology's subject matter is itself calibrated by decision-makers with accumulated economic and political power. This alone should raise questions about whether it is right that some types of evidence are valued more than others—and, indeed, concerns about whose evaluation is counted as mattering (Fine, 2012). Psychologists should reflect more on what type of evidence their field brings to bear on their explorations, and why hierarchies of evidence exist.

They should also reflect on the subject-matter content that psychology has accumulated over its conservative history. That much of psychology is a product of its colonialist past is as uncomfortable as it can be confusing. Colonialism can mean different things to different ('ordinary') people. It can clearly refer to the imposition of Euro-American standards to the formulation of research questions and the gathering and interpretation of data. It can also involve the way Western academia presents its own colonising power structure, in which views, perspectives, and styles of thinking that come from the periphery are subtly—or not so subtly—undermined (Bhatia, 2019). It can result in self-perpetuating dynamics, even among the most well-intentioned psychologists, where individuo-centric approaches to social justice draw attention from wider inequalities (Thrift & Sugarman, 2019), such as depoliticised interventions that assume 'no-fault' forms of racism (Grzanka & Cole, 2021). And, of course, colonialism can manifest as a refusal to acknowledge that psychology has a Whiteness problem, itself a denialism of racism—a 'colour-blind' refusal to acknowledge its enduring impact (Mills, 2007).

The de-privileging of psychology will require an acknowledgement of these historical and cultural dimensions. To deconstruct this colonialist legacy—or, borrowing a phrase from the Indian historian Chakrabarty (2000), to 'provincialise' psychology—the endeavour will require an effort to understand its history and power structures (Clegg, 2022). This will involve an ability to see the wood from the trees, to learn how to deconstruct that which has become normalised in the discipline.

Reclaiming Liberal Values

Perhaps the task that lies ahead is not so much to extinguish any liberal bias within the heart of psychology—or to pander to those (conservatives) who complain about such a thing—but to actively reclaim the concept. Why shouldn't psychology have a 'bias' in favour of a particular set of values? Some will say that any such 'bias' is unacceptable because science is supposed to be 'value-free'. But it is absurd to claim that values are incompatible with good science. Science itself is *reliant* on values.

For example, science is committed to the value of transparency: when scientists refuse to reveal their data or submit their work for peer-review, their practice is said to be *unscientific*. Science is also committed to the value of accuracy: if a scientist wilfully ignores errors in their data, then their work becomes *less scientific*. And science is committed to the value of rigour: if scientists are ambivalent to the principles of methodological rigour—if they use deliberately non-selective samples, ignore the need for control groups, or play fast and loose with measurement accuracy— then this too renders their work *unscientific*. It breaches the value code of science. All scientists know this.

Scientists also adhere to ethical values, where they undertake to respect the rights and dignity of research participants, to have integrity in the handling and reporting of information, and to maintain a basic competence in their research skills, so that standards do not slip over time. Moreover, inherent in their adherence to ethical values is a commitment to values-policing—everyone agrees that scientists who conduct unethical research should be sanctioned, and that their dodgy outputs should be censored. These also are values. All scientists know this too.

The value system of science is conspicuous and frequently discussed. However, it is not dictatorial. Scientists regularly debate the particulars of how values like transparency should be applied in their field, as well as the parameters of permissible value-policing. This is because of another value: free and open discourse. As mentioned above, rational scepticism is central to the scientific method. It is core to science that claims are based on evidence and not on eminence. All claims of fact can, and should, be tested empirically, no matter how esteemed the people who make them. Nobody should get off lightly because of their (presumed) elite status. Scepticism is a necessary condition for good science. It is a *value* that shapes practice.

The type of liberal politics that are supposed to 'bias' psychology encompass a desire for societal justice, a willingness and tolerance for change, support for social intervention, and a respect for diversity of identity. Many would consider such ideas to be extensions of the basic ethics that are applied as standard in scientific research, where the integrity and welfare of humans is held to be sacrosanct. Another orthodox feature of liberal politics is a questioning of contemporary capitalist culture (rather than the unthinking assumption that 'There Is No Alternative'), as well as a scepticism toward all its offshoot philosophies, such as survival-of-the-fittest in-group favouritism and antipathy toward the suffering of outsiders. This questioning stance reflects the integral posture of scientific inquiry, in which statements that claim to be factual are interrogated against evidence.

In real-world settings, liberal politics will likely be concerned with specific cultural practicalities, including the law, the education system, healthcare delivery, social welfare, social media, global commerce, geopolitics, and the military–industrial complex. But when it comes to any of these issues, the type of psychology we have today does more to perpetuate the inequities and injustices inherent in these systems than it does to undo or even discuss them.

The 'liberal bias' claim suggests that psychologists use particular forms of argument, evidence-gathering, data interpretation, and theorising that are skewed by liberalist thinking. But as we have seen throughout this book, this is hardly ever the case. Individuocentrism does not expose a liberal bias: it encourages a form of analysis that actively downplays liberal views of social embeddedness and communitarianism. Reductionism does not expose a liberal bias: it negates explanations of human outcomes that invoke complex systemic causes. Capitalist realism does not expose a liberal bias: it centres all understanding around the one true conservative system of economics. Psychogenic theories of illness and disability do not expose a liberal bias: they attribute responsibility for personal ill-health to victims, instead of to social, economic, or political systems. Psychology's history of embedded assumptions does not expose a liberal bias: it reveals a field that has traditionally depicted out-groups as 'degenerate', women as 'disordered', gender differences as 'inferiorities', and people with disabilities as 'problematic'. Psychology's geographic skew does not expose a liberal bias: it highlights the field's problem of White-centredness and its marginalisation of minority voices. And not even the persistent defence of psychology by self-identifying liberal psychologists exposes a liberal bias in the field: instead, it typifies academic exceptionalism, a tendency toward saviour-ism, and a blindness to flaws that ensures they are left unaddressed.

As such, perhaps it is time to reclaim this idea of 'liberal bias' for psychology. Perhaps it is time to own it. Perhaps it is time to work on cultivating precisely such a bias, where it has too long been absent. Perhaps it is time to acknowledge that the core values of science and human enlightenment—integrity, consistency, solidarity, and fairness—are essential for good psychology. Perhaps it is time to understand that these ideas are not just important because of ethics, they are essential for empirical reasons: a science that constrains itself with individuocentrism, reductionism, capitalist realism, and cultural narrowness will always produce an incomplete account of the human experience. How could psychologists justify tolerating such an outcome?

In the 1950s, the American civil rights leader Martin Luther King Jr. reacted to the pathologising of dissent by declaring that he was 'proud to be maladjusted' (King Jr., 1957). He explained that no healthy person should see themselves as 'well' adjusted to society as it was. King was committed to standing against the many unjust norms embedded in the social order, articulating the view that human beings should never bend their values in response to intimidation, nor internalise the slurs of their aggressors. In much the same way, perhaps it is time for psychology to become 'proud of its liberal orientation'.

It is time for psychologists to commit themselves to challenging—and changing—the comfort-zone conservatism that currently dominates their field. The L-word is not a profanity. Let us declare ourselves proud to use it.

References

Aaronovitch, D. (2014, January 9). I'm sick. It must be one of those syndromes. *The Times*. https://www.thetimes.com/article/im-sick-it-must-be-one-of-those-syndromes-6b39xk3qprc

Abebe, D. S., Lien, L., & Hjelde, K. H. (2014). What we know and don't know about mental health problems among immigrants in Norway. *Journal of Immigrant and Minority Health, 16*, 60–67.

Adams, R. (2021, October 12). Professor says career 'effectively ended' by union's transphobia claims. *The Guardian*. https://www.theguardian.com/education/2021/oct/12/professor-says-career-effectively-ended-by-unions-transphobia-claims

Adams, R. (2025a, March 27). University of Sussex taking legal action over £585,000 free speech fine. *The Guardian*. https://www.theguardian.com/education/2025/mar/27/university-of-sussex-legal-action-free-speech-fine-sasha-roseneil

Adams, R. (2025b, January 15). Ministers to revise university freedom of speech legislation. *The Guardian*. https://www.theguardian.com/education/2025/jan/15/labour-revive-controversial-university-freedom-of-speech-legislation-law

Addley, E. (2023, November 14). What will Esther McVey bring to cabinet as the 'commonsense tsar'? *The Guardian.* https://www.theguardian.com/poli tics/2023/nov/14/what-views-will-esther-mcvey-bring-to-cabinet-as-the-common-sense-tsar

Ahmadi, A. A. (2025, March 8). Trump pulls $400m from Columbia University, saying it failed to protect Jewish students. *BBC News.* https://www.bbc.com/news/articles/c4gp979w055o

Alatas, S. F. (2003). Academic dependency and the global division of labour in the social sciences. *Current Sociology, 51,* 599–613.

al-gharbi, M. (2024). *We have never been woke: The cultural contradictions of a new elite.* Princeton University Press.

Allport, G. W. (1954). The historical background of modern social psychology. In G. Lindzey (Ed.), *Handbook of social psychology* (Vol. 1, pp. 3–56). Addison-Wesley.

Alvesson, M. (2018). Rated: Grandiosity, and functional stupidity. *The Psychologist, 31,* 50–53.

American Association of University Professors. (2024, January). *Statement on political interference in higher education.* https://www.aaup.org/report/state ment-political-interference-higher-education

American Psychological Association. (2014, August 14). *Public information statement in response to the shooting death of Michael Brown.* https://www. apa.org/news/press/releases/2014/08/michael-brown

American Psychological Association. (2020). What do psychology and psychologists offer humanity? *APA Office of International Affairs: Global Insights Newsletter.* https://www.apa.org/international/global-insights/world-needs-psychology

American Psychological Association. (2021, October 29). APA apologizes for long-standing contributions to systemic racism. *American Psychological Association.* https://www.apa.org/news/press/releases/2021/10/apology-systemic-racism

American Psychological Association. (2021). *APA guidelines for psychological practice with sexual minority persons.* APA.

American Psychological Association. (2024, February). APA policy statement on affirming evidence-based inclusive care for transgender, gender diverse, and nonbinary individuals, addressing misinformation, and the role of psychological practice and science. *American Psychological Association.* https://www.apa.org/about/policy/transgender-nonbinary-inclusive-care

Amiri, S. (2022). Unemployment associated with major depression disorder and depressive symptoms: A systematic review and meta-analysis. *International Journal of Occupational Safety and Ergonomics, 28*, 2080–2092.

Amouroux, R., & Zaslawski, N. (2019). 'The damned behaviorist' versus French phenomenologists: Pierre Naville and the French indigenization of Watson's behaviorism. *History of Psychology, 23*, 77–98.

Anderson, C. A., & DeLisi, M. (2011). Implications of global climate change for violence in developed and developing countries. In J. Forgas, A. Kruglanski, & K. Williams (Eds.), *The psychology of social conflict and aggression* (pp. 249–265). Psychology Press.

Aratani, L. (2025, March 29). Two leaders of Harvard's Middle Eastern studies center to step down. *The Guardian.* 5

Associated Press. (1992, October 21). Mattel says it erred; Teen Talk Barbie turns silent on math. *New York Times.* https://www.nytimes.com/1992/10/21/business/company-news-mattel-says-it-erred-teen-talk-barbie-turns-silent-on-math.html

Association of Black Psychologists. (2021, November 24). Official statement — The APA apology: Unacceptable. *Association of Black Psychologists, Inc.* https://abpsi.org/official-statement-the-apa-apology-unacceptable/

Astell, M. (1705). *The Christian religion: As profess'd by a daughter of the Church of England.* R. Wilkin.

Auerbach, R. P., McWhinnie, C. M., Goldfinger, M., Abela, J. R., Zhu, X., & Yao, S. (2010). The cost of materialism in a collectivistic culture: Predicting risky behavior engagement in Chinese adolescents. *Journal of Clinical Child and Adolescent Psychology, 39*, 117–127.

Auguste, E., Nobles, W., & Rowe, D. (2021, November 21). Why the APA's apology for promoting white supremacy falls short. *Think: Opinion, Analysis, Essays.* https://www.nbcnews.com/think/opinion/why-apa-s-apology-promoting-white-supremacy-falls-short-ncna1284229

Bachaud, L., & Johns, J. E. (2023). The use and misuse of evolutionary psychology in online manosphere communities: The case of female mating strategies. *Evolutionary Human Sciences, 5*, e28.

Bachofen, J. J. (1861). *Das Mutterrecht: Eine Untersuchung über die Gynaikokratie der alten Welt nach ihrer religiösen und rechtlichen Natur.* Krais & Hoffmann.

Badham, V. (2024, November 16). Women need to be kept safe from the 'your body, my choice' peddlers. Here's how. *The Guardian.* https://www.theguardian.com/society/commentisfree/2024/nov/17/your-body-my-choice-peddlers-women-self-protection-ntwnfb

Badham, V. (2025, April 30). The 'womansophere' is the latest cultural propaganda assault on young womanhood. Will it work? *The Guardian.* https://www.theguardian.com/commentisfree/2025/apr/30/womanosphere-femosphere-cultural-assault-on-women-ntwnfb

Bahlai, C., Bartlett, L., Burgio, K., Fournier, A., Keiser, C., Poisot, T., & Whitney, K. (2019). Open science isn't always open to all scientists. *American Scientist, 107,* 78–83.

Bailey, J. M. (2023). Ideological bias in sex research. In C. L. Frisby, R. E. Redding, W. T. O'Donohue, & S. O. Lilienfeld (Eds.), *Ideological and political bias in psychology* (pp. 779–803). Springer.

Bajbouj, M., Ta, T. M. T., Hassan, G., & Hahn, E. (2022). Editorial: The nine grand challenges in global mental health. *Frontiers in Psychiatry, 12,* 822299.

Bajwa, S. (2020a). Response from Sarb Bajwa, chief executive of the British Psychological Society. *The Psychologist, 33,* 3.

Bajwa, S. (2020b). Igniting the conversation. *The Psychologist, 33,* 23.

Barnes, S. (2020, July 25). Some local coronavirus outbreaks could be 'mass hysteria', Joint Biosecurity Centre warns. *The Telegraph.* https://www.telegraph.co.uk/news/2020/07/25/local-coronavirus-outbreaks-could-mass-hysteriajoint-biosecurity/

Barr, B., Taylor-Robinson, D., Scott-Samuel, A., McKee, M., & Stuckler, D. (2012). Suicides associated with the 2008–2010 economic recession in England: Time trend analysis. *BMJ, 345,* e5142.

Barry, P. W., Kelley, K., Tan, T., & Finlay, I. (2024). NICE guideline on ME/CFS: Robust advice based on a thorough review of the evidence. *Journal of Neurology, Neurosurgery and Psychiatry, 95,* 671–674.

Bartels, J. M. (2023). Indoctrination in introduction to psychology. *Psychology Learning and Teaching, 22,* 226–236.

Bartle, J. (2019, January 16). The scientific reasons why men are more violent than women. *New York Post.* https://nypost.com/2019/01/16/the-scientific-reasons-why-men-are-more-violent-than-women/

Bartlett, T. (2014, June 23). Replication crisis in psychology research turns ugly and odd. *Chronicle of Higher Education.* http://www.chronicle.com/article/Replication-Crisis-in/147301

Beard, G. M. (1881). *American nervousness: Its causes and consequences a supplement to nervous exhaustion (Neurasthenia).* G. P. Putnam's Sons.

Bettchner, T. M. (2018, May 30). 'When tables speak': On the existence of trans philosophy. *Daily Nous.* https://dailynous.com/2018/05/30/tables-speak-existence-trans-philosophy-guest-talia-mae-bettcher/

Bhatia, S. (2019). Searching for justice in an unequal world: Reframing indigenous psychology as a cultural and political project. *Journal of Theoretical and Philosophical Psychology, 39,* 107–114.

Bhatia, S., & Priya, K. R. (2018). Decolonizing culture: Euro-American psychology and the shaping of neoliberal selves in India. *Theory and Psychology, 28,* 645–668.

Binet, A., & Simon, T. (1916). *The development of intelligence in children* (E. Kite, Trans.). Williams & Wilkins.

Blackless, M., Charuvastra, A., Derryck, A., Fausto-Sterling, A., Lauzanne, K., & Lee, E. (2000). How sexually dimorphic are we? Review and synthesis. *American Journal of Human Biology, 12,* 151–166.

Bourdieu, P. (1993). *Language and symbolic power.* Harvard University Press.

Bourque, F., & Willox, A. C. (2014). Climate change: The next challenge for public mental health? *International Review of Psychiatry, 26,* 415–422.

Bowleg, L., Boone, C. A., Holt, S. L., del Río-gonzález, A. M., & Mbaba, M. (2022). Beyond 'heartfelt condolences': A critical take on mainstream psychology's responses to anti-black police brutality. *American Psychologist, 77,* 362–380.

Bradbury, R., & Cook, J. (2019, April 30). Controversial research fellow Noah Carl dismissed by St Edmund's. *Varsity.* https://www.varsity.co.uk/news/17456

Bradbury, R., & van der Merwe, B. (2019, July 12). Developer who created crowdfunding site for white supremacists set up private limited company for Noah Carl. *Varsity.* https://www.varsity.co.uk/news/17727

Bremmer, I. (2010). *The end of the free market: Who wins the war between states and corporations?* Portfolio.

British Psychological Society. (2024a). Division of Occupational Psychology. *British Psychological Society.* https://www.bps.org.uk/member-networks/division-occupational-psychology

British Psychological Society. (2024b June). Guidelines for psychologists working with gender, sexuality and relationship diversity. *British Psychological Society.* https://explore.bps.org.uk/content/report-guideline/bpsrep.2024.rep129b

British Psychological Society. (2025). What is psychology? *British Psychological Society.* https://www.bps.org.uk/what-psychology

Broudy, H. S. (1954). An analysis of anti-intellectualism. *Educational Theory, 4,* 187–205.

Brown, A. (2024, April 2). The bad science behind Jonathan Haidt's call to regulate social media. *Reason.* https://reason.com/video/2024/04/02/the-bad-science-behind-jonathan-haidts-anti-social-media-crusade/

Brown, A., & Sinclair, A. (2024). *Hate speech frontiers: Exploring the limits of the ordinary and legal concepts.* Cambridge University Press.

Brown, N. J. L., Sokal, A. D., & Friedman, H. L. (2013). The complex dynamics of wishful thinking: The critical positivity ratio. *American Psychologist, 68,* 801–813.

Brozovic, M. (1989). With women in mind. *British Medical Journal, 299,* 689.

Brysbaert, M., & Rastle, K. (2021). *Historical and conceptual issues in psychology* (2nd ed.). Pearson Education.

Buchanan, N. T., Perez, M., Prinstein, M. J., & Thurston, I. B. (2021). Upending racism in psychological science: Strategies to change how science is conducted, reported, reviewed, and disseminated. *American Psychologist, 76,* 1097–1112.

Burns, J. F. (2012, March 9). Norman St. John-Stevas, dapper Tory dismissed by Thatcher, dies at 82. *New York Times.* https://www.nytimes.com/2012/03/10/world/europe/norman-st-john-stevas-tory-dismissed-by-thatcher-dies-at-82.html

Burt, C. (1952). *Intelligence and fertility: The effect of the differential birthrate on inborn mental characteristics.* Cassell and Company.

Buss, D. M., & von Hippel, W. (2018). Psychological barriers to evolutionary psychology: Ideological bias and coalitional adaptations. *Archives of Scientific Psychology, 6,* 148–158.

Button, K. S., Ioannidis, J. P. A., Mokrysz, C., Nosek, B. A., Flint, J., Robinson, E. S. J., & Munafò, M. R. (2013). Power failure: Why small sample size undermines the reliability of neuroscience. *Nature Reviews, Neuroscience, 14,* 365–376.

Bynum, B. (2000). Discarded diagnoses: Drapetomania. *Lancet, 356,* 1615.

Cartwright, S. A. (1851). Report on the diseases and physical peculiarities of the Negro race. *New Orleans Medical and Surgical Journal, 7,* 691–715.

Casey, P. R. (2010). Abortion among young women and subsequent life outcomes. *Best Practice and Research: Clinical Obstetrics and Gynaecology, 24,* 491–502.

Castelnuovo-Tedesco, P. (1962). Emotional antecedents of perforation of ulcers of the stomach and duodenum. *Psychosomatic Medicine, 24,* 398–416.

Cattell, R. B. (1972). *A new morality from science: Beyondism.* Pergamon.

Cattell, R. B. (1987). *Beyondism: Religion from science.* Praeger.

Chakrabarty, D. (2000). *Provincializing Europe: Postcolonial thought and historical difference*. Princeton University Press.

Cheek, J. (2008). Healthism: A new conservatism? *Qualitative Health Research, 18*, 974–982.

Chesterton, G. K. (1922). *Eugenics and other evils*. Cassell and Company.

Chiao, J. Y. (2009). Cultural neuroscience: A once and future discipline. *Progress in Brain Research, 178*, 287–304.

Chivers, T. (2014, August 29). Susan Greenfield: 'I'm not scaremongering'. *The Telegraph*. https://www.telegraph.co.uk/culture/books/11050871/Susan-Greenfield-Im-not-scaremongering.html

Chomsky, N. (1967, February 23). The responsibility of intellectuals. *New York Review of Books*. https://www.nybooks.com/articles/1967/02/23/a-special-supplement-the-responsibility-of-intelle/

Christakis, N. A. [@NAChristakis]. (2024, May 3). *There do seem to be elements of MPI ("mass psychogenic illness"—Formerly known as "epidemic hysteria") in the present student protests (as in most mass movements!).* @connected_book [Post]. X. https://x.com/NAChristakis/status/1786355746457698553

Christoph [@Halalcoholism] (2018, October 1). *Yo, remind me again which side disagrees with scientific consensus on evolution and climate change* [Post]. Twitter. https://x.com/Halalcoholism/status/1046671310737235968

Cipriani, E., Frumento, S., Gemignani, A., & Menicucci, D. (2024). Personality traits and climate change denial, concern, and proactivity: A systematic review and meta-analysis. *Journal of Environmental Psychology, 95*, 102277.

Clarence-Smith, L. (2023, February 6). Free speech law 'must let cancelled academics sue woke universities'. *The Telegraph*. https://www.telegraph.co.uk/politics/2023/02/06/free-speech-law-must-let-cancelled-academics-sue-woke-universities/

Clark, E. H. (1873). *Sex in education; or, a fair chance for the girls*. James R. Osgood and Company.

Clark, L. P. (1926). A further contribution to the psychology of the essential epileptic. *Journal of Nervous and Mental Disease, 63*, 575–585.

Clark, M. (2023, November 27). Selma Blair says 'older male doctors' misdiagnosed her multiple sclerosis as menstrual issues. *Independent*. https://www.independent.co.uk/life-style/health-and-families/selma-blair-ms-misdiagnosis-symptoms-b2454317.html

Clearing House for Postgraduate Courses in Clinical Psychology (2024, September 3). Equal opportunities data: 2023 entry. *Equal Opportunities.* https://www.clearing-house.org.uk/system/files/2024-05/Equal%20opportu nities%20data%20for%202023%20entry.pdf

Clegg, J. W. (2022). *Good science: Psychological inquiry as everyday moral practice.* Cambridge University Press.

Clifford, E. (2020). *The war on disabled people: Capitalism, welfare and the making of a human catastrophe.* Zed Books.

Coates, S., & Sylvester, R. (2016, July 9). Being a mother gives me edge on MayLeadsom. *The Times.* https://www.thetimes.com/article/being-a-mother-gives-me-edge-on-may-leadsom-0t7bbm29x

Cohen, L., & Lansing, A. H. (2021). The tend and befriend theory of stress: Understanding the biological, evolutionary, and psychosocial aspects of the female stress response. In H. Hazlett-Stevens (Ed.), *Biopsychosocial factors of stress, and mindfulness for stress reduction* (pp. 67–81). Springer.

Cohen, S. (1972). *Folk devils and moral panics: The creation of the mods and rockers.* MacGibbon and Kee.

Collin, L., Reisner, S. L., Tangpricha, V., & Goodman, M. (2016). Prevalence of transgender depends on the 'case' definition: A systematic review. *Journal of Sexual Medicine, 13,* 613–626.

Collins, A. F. (1999). The enduring appeal of physiognomy: Physical appearance as a sign of temperament, character, and intelligence. *History of Psychology, 2,* 251–276.

Colvile, R. (2016, July 20). Spot the WEIRDo. *Aeon.* https://aeon.co/essays/american-undergrads-are-too-weird-to-stand-for-all-humanity

Confessore, N. (2024, January 2). How a proxy fight over campus politics brought down Harvard's president. *New York Times.* https://www.nytimes.com/2024/01/02/us/harvard-president-campus-antisemitism-conservatives.html

Constantine, M. G. (2007). Racial microaggressions against African American clients in cross-racial counseling relationships. *Journal of Counseling Psychology, 54,* 1–16.

Coon, D. J. (1992). Testing the limits of sense and science: American experimental psychologists combat spiritualism, 1880–1920. *American Psychologist, 47,* 143–151.

Cooperrider, K. (2019, January 23). What happens to cognitive diversity when everyone is more WEIRD? *Aeon.* https://aeon.co/ideas/what-happens-to-cognitive-diversity-when-everyone-is-more-weird

Cosgrove, L., & Karter, J. M. (2018). The poison in the cure: Neoliberalism and contemporary movements in mental health. *Theory & Psychology, 28,* 669–683.

Costello, T. H. (2023). The conundrum of measuring authoritarianism: A case study in political bias. In C. L. Frisby, R. E. Redding, W. T. O'Donohue, & S. O. Lilienfeld (Eds.), *Ideological and political bias in psychology* (pp. 585–601). Springer.

Craig, R., Pelosi, A., & Tourish, D. (2020). Research misconduct complaints and institutional logics: The case of Hans Eysenck and the British Psychological Society. *Journal of Health Psychology, 26,* 296–311.

Crawford, J. T., & Jussim, L. (Eds.). (2018). *The politics of social psychology.* Psychology Press.

Crist, E. (2020). Let earth rebound! Conservation's new imperative. In H. Kopnina,& H. Washington (Eds.), *Conservation: Integrating social and ecological justice* (pp. 201–218). Springer.

Crohn, B. B. (1947). Peptic ulcer as a psychosomatic disease. *Surgical Clinics of North America, 27,* 309–314.

Cruz-Hernandez, A., Roney, A., Goswami, D. G., Tewari-Singh, N., & Brown, J. M. (2022). A review of chemical warfare agents linked to respiratory and neurological effects experienced in Gulf War Illness. *Inhalation Toxicology, 34,* 412–432.

Cummings Center for the History of Psychology at the University of Akron (2021, October). Historical chronology. *American Psychological Association.* https://www.apa.org/about/apa/addressing-racism/historical-chronology

Curtis, D., & Balloux, F. (2020). Topical ethical issues in the publication of human genetics research. *Annals of Human Genetics, 84,* 313–314.

Danziger, K. (2013). Psychology and its history. *Theory & Psychology, 23,* 829–839.

Daum, M. (2022, October 4). What a conservative therapist thinks about politics and mental health. *New York Times.* https://www.nytimes.com/2022/10/04/opinion/us-conservative-therapy-politics.html

Davies, A. (2023, July 20). Jim Davidson blasts 'nutcase' just stop oil protestors after violent clashes with public. *GB News.* https://www.gbnews.com/celebrity/jim-davidson-just-stop-oil-dan-wootton

Davis, B. (2022, October 19). Suella Braverman blames 'Guardian-reading tofu-eating wokerati' for disruptive protests. *The Standard.* https://www.standard.co.uk/news/uk/suella-braverman-guardian-tofu-eating-wokerati-extinction-rebellion-just-stop-oil-protest-action-b1033677.html

DeAngelis, T., & Andoh, E. (2022). Confronting past wrongs and building an equitable future. *Monitor on Psychology, 53*, 22.

Deary, V., & Chalder, T. (2010). Personality and perfectionism in chronic fatigue syndrome: A closer look. *Psychology & Health, 25*, 466–475.

Deckop, J. R., Giacalone, R. A., & Jurkiewicz, C. L. (2015). Materialism and workplace behaviors: Does wanting more result in less? *Social Indicators Research, 121*, 787–803.

Deckop, J. R., Jurkiewicz, C. L., & Giacalone, R. A. (2010). Effects of materialism on work-related personal well-being. *Human Relations, 63*, 1007–1030.

Degerman, D. (2019a, April 3). Brexit anxiety shouldn't be over-medicalised: It is fuelling real political engagement. *The Conversation.* https://theconversation.com/brexit-anxiety-shouldnt-be-over-medicalised-it-is-fuellingreal-political-engagement-114664

Degerman, D. (2019b). Brexit anxiety: A case study in the medicalization of dissent. *Critical Review of International Social and Political Philosophy, 22*, 823–840.

DeJesus, J. M., Callanan, M. A., Solis, G., & Gelman, S. A. (2019). Generic language in scientific communication. *Proceedings of the National Academy of Sciences, 116*, 18370–18377.

Delaney, H. D., Miller, W. R., & Bisonó, A. M. (2013). Religiosity and spirituality among psychologists: A survey of clinician members of the American Psychological Association. *Spirituality in Clinical Practice, 1*, 95–106.

DeLuca, J. S., & Yanos, P. T. (2016). Managing the terror of a dangerous world: Political attitudes as predictors of mental health stigma. *International Journal of Social Psychiatry, 62*, 21–30.

Deutsch, A. (1944). The first U.S. Census of the Insane (1840) and its use as pro-slavery propaganda. *Bulletin of the History of Medicine, 15*, 469–482.

Dewey, J. (1884). The new psychology. *Andover Review, 2*, 278–289.

Dewsbury, S. (2016). Uncanny interpretations of class and authority in *Monty Python's Flying Circus.* In R. W. Wolny & S. Nicieja (Eds.), *The outlandish, uncanny, bizarre: Culture literature philosophy* (pp. 145–159). Wyższej Szkoły Filologicznej we Wrocławiu.

Dhont, K., Hodson, G., Leite, A. C., & Salmen, A. (2019). The psychology of speciesism. In K. Dhont, & G. Hodson (Eds.), *Why we love and exploit animals: Bridging insights from academia and advocacy* (pp. 29–49). Routledge.

Dittmar, H., Bond, R., Hurst, M., & Kasser, T. (2014). The relationship between materialism and personal well-being: A meta-analysis. *Journal of Personality and Social Psychology, 107*, 879–924.

Dormandy, T. (2000). *The white death: A history of tuberculosis.* New York University Press.

Dragioti, E., Dimoliatis, I., Fountoulakis, K. N., & Evangelou, E. (2015). A systematic appraisal of allegiance effect in randomized controlled trials of psychotherapy. *Annals of General Psychiatry, 14*, 25.

du Pré, H., & du Pré, P. (1997). *A genius in the family: An intimate Memoir of Jacqueline du Pré.* Heinemann.

Duarte, J. L., Crawford, J. T., Stern, C., Haidt, J., Jussim, L., & Tetlock, P. E. (2015). Political diversity will improve social psychological science. *Behavioral and Brain Sciences, 38*, e130.

Duckworth, A. L., Steen, T. A., & Seligman, M. E. (2005). Positive psychology in clinical practice. *Annual Review of Clinical Psychology, 1*, 629–651.

Dunn, S. M., & Turtle, J. R. (1981). The myth of the diabetic personality. *Diabete Care, 4*, 640–646.

Durkin, E. (2019, January 13). DNA scientist James Watson stripped of honors over views on race. *The Guardian.* https://www.theguardian.com/world/2019/jan/13/james-watson-scientist-honors-stripped-reprehensible-race-comments

Durrheim, K. (2024). Conversational silencing of racism in psychological science: Toward decolonization in practice. *Perspectives on Psychological Science, 19*, 244–257.

Ecklund, E. H., & Scheitle, C. P. (2007). Religion among academic scientists: Distinctions, disciplines, and demographics. *Social Problems, 54*, 289–307.

Eddo-Lodge, R. (2017). *Why I'm no longer talking to white people about race.* Bloomsbury.

Edwards, J. (2017). PACE team response shows a disregard for the principles of science. *Journal of Health Psychology, 22*, 1155–1158.

Ehrenreich, B. (2009). *Smile or die: How positive thinking fooled America and the world.* Granta.

Eigenbrodt, A. K., Ashina, H., Khan, S., Diener, H. C., Mitsikostas, D. D., Sinclair, A. J., Pozo-Rosich, P., Martelletti, P., Ducros, A., Lantéri-Minet, M., Braschinsky, M., Del Rio, M. S., Daniel, O., Özge, A., Mammadbayli, A., Arons, M., Skorobogatykh, K., Romanenko, V., Terwindt, G. M., ... Ashina, M., et al. (2021). Diagnosis and management of migraine in ten steps. *Nature Reviews Neurology, 17*, 501–514.

Enserink, M. (2011, September 7). Dutch university sacks social psychologist over faked data. *Science.* https://www.science.org/content/article/dutch-university-sacks-social-psychologist-over-faked-data

Eysenck, H. J. (1973). *The inequality of man.* EdITS Publishers.

Eysenck, H. J. (1991). *Smoking, personality, and stress: Psychosocial factors in the prevention of cancer and coronary heart disease.* Springer-Verlag.

Eysenck, H. J., & Grossarth-Maticek, R. (1991). Creative novation behaviour therapy as a prophylactic treatment for cancer and coronary heart disease: II. Effects of treatment. *Behaviour Research and Therapy, 29,* 17–31.

Fadus, M. C., Ginsburg, K. R., Sobowale, K., Halliday-Boykins, C. A., Bryant, B. E., Gray, K. M., & Squeglia, L. M. (2020). Unconscious bias and the diagnosis of disruptive behavior disorders and ADHD in African American and Hispanic youth. *Academic Psychiatry, 44,* 95–102.

Falk, J. H., & Dierking, L. D. (2002). *Lessons without limit: How free-choice learning is transforming education.* AltaMira Press.

Fanon, F. (1952). *Peau Noire, Masques Blancs.* Éditions du Seuil.

Fanon, F. (1961). *Les damnés de la terre.* Éditions François Maspero.

Farris, S. G., Thomas, J. G., Kibbey, M. M., Pavlovic, J. M., Steffen, K. J., & Bond, D. S. (2020). Treatment effects on pain catastrophizing and cutaneous allodynia symptoms in women with migraine and overweight/obesity. *Health Psychology, 39,* 927–933.

Federal Bureau of Investigation (2023, March 13). FBI releases supplemental 2021 hate crime statistics. https://www.fbi.gov/news/press-releases/fbi-releases-supplemental-2021-hate-crime-statistics

Ferguson, E. (2021, February 14). 'War on WOKE!': Ministers to crack down on universities with 'free speech champion' plans. *Sunday Express.* https://www.express.co.uk/news/politics/1397626/cancel-culture-woke-uk-university-free-speech-champion-oliver-dowden-british-history

Ferguson, N. (2025, February 28). How EDI, cancel culture and bad boards are killing our universities. *The Times.* https://www.thetimes.com/comment/columnists/article/edi-cancel-culture-bad-boards-killing-universities-gd9rzr0rb

Fernald, A. (2010). Getting beyond the 'convenience sample' in research on early cognitive development. *Behavioral and Brain Sciences, 33,* 91–92.

Fernández-Ríos, L., & Novo, M. (2012). Positive psychology: *Zeitgeist* (or spirit of the times) or ignorance (or disinformation) of history? *International Journal of Clinical and Health Psychology, 12,* 333–344.

Fernández-Ríos, L., & Vilariño, M. (2016). Myths of positive psychology: Deceptive manoeuvres and pseudoscience. *Papeles del Psicólogo, 37*, 134–142.

Fine, M. (2012). Troubling calls for evidence: A critical race, class and gender analysis of whose evidence counts. *Feminism & Psychology, 22*, 3–19.

Fine, M. (2016). Just methods in revolting times. *Qualitative Research in Psychology, 13*, 347–365.

Finkel, E. J., Bail, C. A., Cikara, M., Ditto, P. H., Iyengar, S., Klar, S., Mason, L., McGrath, M. C., Nyhan, B., Rand, D. G., Skitka, L. J., Tucker, J. A., Van Bavel, J. J., Wang, C. S., & Druckman, J. N. (2020). Political sectarianism in America. *Policy Forum, 370*, 533–536.

Fisher, M. (2009). *Capitalist realism: Is there no alternative?* O Books.

Fordtran, J. S. (1973). The psychosomatic theory of peptic ulcer. In M. H. Sleisenger & J. S. Fordtran (Eds.), *Gastrointestinal disease* (pp. 163–173). W. B. Saunders.

Foster, A. (2022, September 3). These annoying eco lunatics really need to grow up, says Dame Arlene Foster. *Daily Express*. https://www.express.co.uk/comment/expresscomment/1664073/eco-warriors-extinction-rebellion

Foucault, M. (1961). *Folie et déraison: Histoire de la folie À l'âge classique.* Librairie Plon.

Frawley, A. (2015). Happiness research: A review of critiques. *Sociology Compass, 9*, 62–77.

Freedman, S. (2017, March 21). 'Do mad people get endometriosis or does endo make you mad?' *The Guardian*. https://www.theguardian.com/commentisfree/2017/mar/22/do-mad-people-get-endometriosis-or-does-endo-make-you-mad

Freire, P. (1970). *Pedagogia Do Oprimido.* Afrontamento.

Frisby, C. L., Redding, R. E., O'Donohue, W. T., & Lilienfeld, S. O.(Eds.). (2003). *Ideological and political bias in psychology.* Springer.

Fuchs, A. H. (2000). Contributions of American mental philosophers to psychology in the United States. *History of Psychology, 3*, 3–19.

Fuller, R. C. (2006). American psychology and the religious imagination. *Journal of the History of the Behavioral Sciences, 42*, 221–235.

Furnham, A., & Cheng, A. (2023). Childhood onset of migraine, gender, psychological distress and locus of control as predictors of migraine in adulthood. *Psychology, Health & Medicine, 28*, 2045–2057.

Gabbatiss, J. (2019, January 13). James Watson: The most controversial statements made by the father of DNA. *The Independent.* https://www. independent.co.uk/news/science/james-watson-racism-sexism-dna-race-int elligence-genetics-double-helix-a8725556.html

Galton, F. (1853). *Narrative of an explorer in tropical South Africa.* Ward, Lock and Co.

Garb, H. N. (2021). Race bias and gender bias in the diagnosis of psychological disorders. *Clinical Psychology Review., 90,* 102087.

Garcia, E. (2025, January 13). As wildfires rage, this conservative attacks sign language interpreters sharing news. *MSNBC.* https://www.msnbc.com/opi nion/msnbc-opinion/california-wildfires-deaths-asl-chris-rufo-rcna187341

García-Alandete, J. (2024). Magda Arnold's understanding of the human person: Thomistic personalism, psychophysical unity of the person, integration of personality, and transcendence. *History of Psychology, 27,* 159–177.

Garðarsdóttir, R. B., Dittmar, H., & Aspinall, C. (2009). It's not the money, it's the quest for a happier self: The role of happiness and success motives in the link between financial goals and subjective well-being. *Journal of Social and Clinical Psychology, 28,* 1100–1127.

Gattuso, P. (2023, September 8). Harvard university awarded 'abysmal' speech-climate rating. *National Review.* https://www.nationalreview.com/corner/ harvard-university-awarded-abysmal-speech-climate-rating/

GBD. (2018). Global, regional, and national incidence, prevalence, and years lived with disability for 354 diseases and injuries for 195 countries and territories, 1990-2017: A systematic analysis for the Global Burden of Disease Study 2017. *Lancet, 392,* 1789–1858.

Gelman, A. (2019, August 15). Replication police methodological terrorism stasi nudge shoot the messenger wtf. *Statistical Modeling, Causal Inference, and Social Science.* https://statmodeling.stat.columbia.edu/2019/08/15/replica tion-police-methodological-terrorism-stasi-nudge-shoot-the-messenger-wtf/

Gessen, M. (2020). *Surviving autocracy.* Granta.

Gilbert, D. T., King, G., Pettigrew, S., & Wilson, T. D. (2016). Comment on 'Estimating the reproducibility of psychological science'. *Science, 351,* 1037–a.

Glass, B. (1986). Geneticists embattled: Their stand against rampant eugenics and racism in America during the 1920s and 1930s. *Proceedings of the American Philosophical Society, 130,* 130–154.

Goldberg, E. (2015, August 17). Trump Derangement Syndrome. *American Spectator.* https://spectator.org/63786_trump-derangement-syndrome/

Goldenberg, S. (2005, January 18). Why women are poor at science, by Harvard president. *The Guardian.* https://www.theguardian.com/science/2005/jan/18/educationsgendergap.genderissues

Gómez, J. M., Caño, A., & Baltes, B. B. (2021). Who are we missing? Examining the graduate record examination quantitative score as a barrier to admission into psychology doctoral programs for capable ethnic minorities. *Training and Education in Professional Psychology, 15,* 211–218.

Gondra, J. M. (2013). A psychology of liberation for Central America: The unfinished work of Ignacio Martín-Baró (1942–1989). *Spanish Journal of Psychology, 16,* 1–17.

Goozee, H. (2021). Decolonizing trauma with Frantz Fanon. *International Political Sociology, 15,* 102–120.

Goudsmit, E. M. (1994). All in her mind! Stereotypic views and the psychologisation of women's illness. In S. Wilkinson & C. Kitzinger (Eds.), *Women and health: Feminist perspectives* (pp. 7–12). Taylor & Francis.

Graham, P. J., Rutter, M. L., Yule, W., & Pless, I. B. (1967). Childhood asthma: A psychosomatic disorder? Some epidemiological considerations. *British Journal of Preventive and Social Medicine, 21,* 78–85.

Graham, S. (2024). *Rebel bodies: A guide to the gender health gap revolution.* Green Tree.

Green, G., & Gilbertson, J. (2008). *Warm front better health: Health impact evaluation of the warm front scheme.* Centre for Regional, Economic and Social Research, Sheffield Hallam University.

Griffiths, S. (2021, October 10). Kathleen Stock, the Sussex University professor in trans row, urged to get bodyguards. https://www.thetimes.co.uk/article/kathleen-stock-the-sussex-university-professor-in-trans-row-urged-to-get-bodyguards-2khmgzk98

Griffiths, V. (2025, March 14). Labour's handling of immigration is 'performative cruelty'. *Sussex Bylines.* https://sussexbylines.co.uk/politics/human-rights/labours-handling-of-immigration-is-performative-cruelty/

Gross, P. R., & Levitt, N. (1994). *Higher superstition: The academic left and its quarrels with science.* Johns Hopkins University Press.

Grossarth-Maticek, R., Eysenck, H. J., & Vetter, H. (1988a). Personality type, smoking habit and their interaction as predictors of cancer and coronary heart disease. *Personality and Individual Differences, 9,* 479–495.

Grossarth-Maticek, R., Eysenck, H. J., & Vetter, H. (1988b). Antismoking attitudes and general prejudice: An empirical study. *Perceptual and Motor Skills, 66,* 927–931.

Grzanka, P. R., & Cole, E. R. (2021). An argument for bad psychology: Disciplinary disruption, public engagement, and social transformation. *American Psychologist, 76*, 1334–1345.

Gurfinkel, J. (2024, August 14). Dx machina. *Jenka's Substack*. https://jenka. substack.com/p/dx-machina

Guthrie, R. V. (2004). *Even the rat was white: A historical view of psychology*. Pearson.

Haidt, J. (2016). Why concepts creep to the left. *Psychological Inquiry, 27*, 40–45.

Haidt, J. (2024a). *The anxious generation: How the great rewiring of childhood is causing an epidemic of mental illness*. Penguin.

Haidt, J. (2024b, March 24). Generation anxiety: Smartphones have created a gen Z mental health crisisbut there are ways to fix it. https://www. theguardian.com/books/2024/mar/24/the-anxious-generation-jonathan-haidt-book-extract-instagram-tiktok-smartphones-social-media-screens

Haley, R. W., Kramer, G., Xiao, J., Dever, J. A., & Teiber, J. (2022). Evaluation of a gene-environmental interaction of PON1 and low-level nerve agent exposure with Gulf War Illness: A prevalence case-control study drawn from the US military health survey's national population sample. *Environmental Health Perspectives, 130*, 57001.

Hall, G. S. (1885). The new psychology. *Andover Review, 3*, 242–259.

Hall, G. S. (1894). The new psychology as a basis of education. *Forum, 17*, 710–720.

Hall, G. S. (1901). The new psychology. *Harper's Monthly, 103*, 727–732.

Hall, G. S. (1906). The question of coeducation. *Munsey's Magazine, 34*, 588–592.

Hall, G. S. (1911). Eugenics: Its ideals and what it is going to do. *Religious Education, 6*, 152–159.

Hall, R. (2023, July 11). ME/CFS guidance that discourages exercise is flawed, say researchers. *The Guardian*. https://www.theguardian.com/society/2023/jul/11/chronic-fatigue-guidance-discouraging-exercise-is-flawed-say-researchers

Harris, J. (2024, December 29). The culture wars are coming for children with special needs — Labour must tread carefully. *The Guardian*. https://www. theguardian.com/commentisfree/2024/dec/29/children-special-needs-councils-labour-send

Hart, B., & Spearman, C. (1912). General ability, its existence and nature. *British Journal of Psychology, 5*, 51–84.

Hassett, A. L., Radvanski, D. C., Buyske, S., Savage, S. V., & Sigal, L. H. (2009). Psychiatric comorbidity and other psychological factors in patients with 'chronic Lyme disease'. *American Journal of Medicine, 122,* 843–850.

Heffer, T., Good, M., Daly, O., MacDonell, E., & Willoughby, T. (2019). The longitudinal association between social-media use and depressive symptoms among adolescents and young adults: An empirical reply to Twenge et al. (2018). *Clinical Psychology Science, 7,* 462–470.

Heltzel, G., & Laurin, K. (2020). Polarization in America: Two possible futures. *Current Opinion in Behavioral Sciences, 34,* 179–184.

Hemel, D. J. (2005, January 14). Summers' comments on women and science draw ire. *Harvard Crimson.* https://www.thecrimson.com/article/2005/1/14/summers-comments-on-women-and-science/

Henrich, J., Heine, S. J., & Norenzayan, A. (2010). The weirdest people in the world? *Behavioral and Brain Sciences, 33,* 61–83.

Hilbig, B. E., & Moshagen, M. (2015). A predominance of self-identified democrats is no evidence of a leftward bias. *Behavioral and Brain Sciences, 38,* e146.

Hicks, J. W. (2004). Ethnicity, race, and forensic psychiatry: Are we color-blind? *Journal of the American Academy of Psychiatry and the Law, 32,* 21–33.

Higgins, T. (2023, December 23). Why Elon Musk won't stop talking about a 'woke mind virus'. *Wall Street Journal.* https://www.wsj.com/tech/elon-musk-woke-mind-virus-41576aa6

Hinsliff, G. (2021, November 3). The battle for Stonewall: The LGBT charity and the UK's gender wars. *New Statesman.* https://www.newstatesman.com/politics/2021/11/the-battle-for-stonewall-the-lgbt-charity-and-the-uks-gender-wars

Hitchens, P. (2025, March 12). Dyslexia doesn't exist. It's a made-up affliction that's become a multi-million pound industry around children who haven't been taught to read. *Daily Mail.* https://www.dailymail.co.uk/debate/article-14491505/PETER-HITCHENS-Dyslexia-does-not-exist.html

Hoffman, L. (2025, March 26). 'Trump derangement syndrome' and the Goldwater rule for psychiatrists. *The Guardian.* https://www.theguardian.com/us-news/2025/mar/26/trump-derangement-syndrome-and-the-goldwater-rule-for-psychiatrists

Holcomb, J. L. (2016). *Moral commerce: Quakers and the transatlantic boycott of the slave labor economy.* Cornell University Press.

Holmes, M. E. (2020). Anglo-American psychology in the Cold War. *Oxford Research Encyclopedia of Psychology.* March 31. https://doi.org/10.1093/acrefore/9780190236557.013.610

Holowinsky, I. Z. (1985). Soviet psychology and its view of American behaviorism. *Psychological Reports.*, *56*, 803–810.

Honeycutt, N., & Jussim, L. (2023). Political bias in the social sciences: A critical, theoretical, and empirical review. In C. L. Frisby, R. E. Redding, W. T. O'Donohue, & S. O. Lilienfeld (Eds.), *Ideological and political bias in psychology* (pp. 97–146). Springer.

Houwen, J., Lucassen, P. L., Dongelmans, S., Stappers, H. W., Assendelft, W. J., van Dulmen, S., & Olde Hartman, T. C. (2020). Medically unexplained symptoms: Time to and triggers for diagnosis in primary care consultations. *British Journal of General Practice*, *70*, e86–e94.

Hughes, B. (2021, August 10). Journalists covering ME/CFS: Don't ask about the new NICE guideline, ask about the old one. *The Science Bit*. https://thesciencebit.net/2021/08/10/journalists-covering-me-cfs-dont-ask-about-the-new-nice-guideline-ask-about-the-old-one/

Hughes, B. (2023). *A conceptual history of psychology: The mind through time*. Bloomsbury.

Hughes, B. M. (2016). *Rethinking psychology: Good science, bad science, pseudoscience*. Palgrave.

Hughes, B. M. (2018a). *Psychology in crisis*. Palgrave.

Hughes, B. M. (2018b). Does psychology face an exaggeration crisis? *The Psychologist*, *31*, 8–10.

Hughes, B. M. (2019). *The psychology of Brexit: From psychodrama to behavioural science*. Palgrave Macmillan.

Human Rights Campaign Foundation. (2023, November). *The epidemic of violence against the transgender and gender non-conforming community in the United States*. The 2023 report. https://reports.hrc.org/an-epidemic-of-violence-2023

Hyde, J. S. (2005). The gender similarities hypothesis. *American Psychologist*, *60*, 581–592.

Hyde, J. S. (2014). Gender similarities and differences. *Annual Review of Psychology*, *65*, 373–398.

Hyde, J. S. (2018). Gender similarities. In C. B. Travis, J. W. White, A. Rutherford, W. S. Williams, S. L. Cook, & K. F. Wyche (Eds.), *APA handbook of the psychology of women: History, theory, and battlegrounds* (pp. 129–143). American Psychological Association.

Hyde, J. S., Bigler, R. S., Joel, D., Tate, C. C., & van Anders, S. M. (2019). The future of sex and gender in psychology: Five challenges to the gender binary. *American Psychologist*, *74*, 171–193.

Hyde, J. S., Lindberg, S. M., Linn, M. C., Ellis, A. B., & Williams, C. C. (2008). Gender similarities characterize math performance. *Science*, *321*, 494–495.

Iemmi, V., Bantjes, J., Coast, E., Channer, K., Leone, T., McDaid, D., Palfreyman, A., Stephens, B., & Lund, C. (2016). Suicide and poverty in low-income and middle-income countries: A systematic review. *The Lancet Psychiatry*, *3*, 774–783.

IJzerman, H., Lewis Jr, N. A., Przybylski, A. K., Weinstein, N., DeBruine, L., Ritchie, S. J., Vazire, S., Forscher, P. S., Morey, R. D., Ivory, J. D., & Anvari, F. (2020). Use caution when applying behavioural science to policy. *Nature Human Behaviour*, *4*, 1092–1094.

Illingworth, C. F. W. (1956). Inborn and extraneous factors in the aetiology of peptic ulcer. *Journal of the Royal College of Surgeons of Edinburgh*, *2*, 14–23.

Ilyes, E. (2020). Psychology's eugenic history and the invention of intellectual disability. *Social and Personality Psychology Compass*, *14*, e12537.

Inbar, Y., & Lammers, J. (2012). Political diversity in social and personality psychology. *Perspectives on Psychological Science*, *7*, 496–503.

Ioannidis, J. P. (2005). Why most published research findings are false. *PLoS Medicine*, *2*, e124.

Jackson, B. (2014). *The inspirational atheist: Wise words on the wonder and meaning of life*. Plume.

Jahoda, G. (2007). *A history of social psychology: From the eighteenth-century enlightenment to the second world war*. Cambridge University Press.

James, J. E. (2016). *The health of populations: Beyond medicine*. Academic Press.

James, W. (1890). *The principles of psychology* (Vol. I). Henry Holt and Company.

Jeffreys, B. (2025, March 25). *University of Sussex fined £585k in transgender free speech row*. BBC News. https://www.bbc.com/news/articles/cn9vr4vjzgqo

Jelliffe, S. E. (1921). Emotional and psychological factors in multiple sclerosis. *Proceedings of the Association for Research in Nervous and Mental Disease*, *2*, 82–95.

Jing, F., Li, Z., Qiao, S., Ning, H., Zhou, S., & Li, X. (2023). Association between immigrant concentration and mental health service utilization in the United States over time: A geospatial big data analysis. *Health & Place*, *83*, 103055.

Johnson, A., Shapiro, L. B., & Franz, A. (1947). Preliminary report on a psychosomatic study of rheumatoid arthritis. *Psychosomatic Medicine, 9*, 295–300.

Jonason, P. K., & Schmitt, D. P. (2016). Quantifying common criticisms of evolutionary psychology. *Evolutionary Psychological Science, 2*, 177–188.

Jones, D., & Elcock, J. (2001). *History and theories of psychology: A critical perspective.* Hodder Education.

Jussim, L. (2019, December 27). The threat to academic freedom . . . from academics. *PsychRabble.* https://psychrabble.medium.com/the-threat-to-academic-freedom-from-academics-4685b1705794

Kahneman, D. (2012). *Thinking, fast and slow.* Penguin.

Kanceljak, D., & Calia, C. (2023). Diversity and inclusion in UK psychology: A nationwide survey. *Clinical Psychology Forum, 369*, 15–29.

Kane, J. M., & Mertz, J. E. (2012). Debunking myths about gender and mathematics performance. *Notices of the American Mathematical Society, 59*, 10–21.

Karakurum, B., Soylu, O., Karataş, M., Giray, S., Tan, M., Arlier, Z., & Benli, S. (2004). Personality, depression, and anxiety as risk factors for chronic migraine. *International Journal of Neuroscience, 114*, 1391–1399.

Karpov, Y. V. (2024). Elite universities: Incubators of leftist ideology. *Academic Questions, 37*, 28–36.

Kasser, T. (2016). Materialistic values and goals. *Annual Review of Psychology, 67*, 489–514.

Kasser, T., Rosenblum, K. L., Sameroff, A. J., Deci, E. L., Niemiec, C. P., Ryan, R. M., Árnadóttir, O., Bond, R., Dittmar, H., Dungan, N., & Hawks, S. (2014). Changes in materialism, changes in psychological well-being: Evidence from three longitudinal studies and an intervention experiment. *Motivation and Emotion, 38*, 1–22.

Kaufmann, E. (2021, March 1). Academic freedom in crisis: Punishment, political discrimination, and self-censorship. *Center for the Study of Partisanship and Ideology.* https://www.cspicenter.com/p/academic-freedom-in-crisis-punishment

Keenan, H. T., Marshall, S. W., Mocera, M. A., & Runyan, D. K. (2004). Increased incidence of inflicted traumatic brain injury in children after a natural disaster. *American Journal of Preventive Medicine., 26*, 189–193.

Kendi, I. X. (2023, March 23). The crisis of the intellectuals. *The Atlantic.* https://www.theatlantic.com/ideas/archive/2023/03/intellectualism-crisis-american-racism/673480/

Kentish, B. (2019, March 27). Conservative MP condemned for repeating far-right, antisemitic conspiracy theory about 'cultural Marxism'. *The Independent.* https://www.independent.co.uk/news/uk/politics/cultural-marxism-tory-suella-braverman-conservatives-antisemitism-a8841581.html

Khafaie, M. A., Sayyah, M., & Rahim, F. (2019). Extreme pollution, climate change, and depression. *Environmental Science and Pollution Research, 26,* 22103–22105.

Kim, K. M., Kim, D., & Chung, U. S. (2020). Investigation of the trend in adolescent mental health and its related social factors: A multi-year cross-sectional study for 13 years. *International Journal of Environmental Research & Public Health, 17,* 5405.

King Jr., M. L. (1957[1986]). The power of nonviolence. In J. M. Washington (Ed.), *A testament of hope: The essential writings of Martin Luther King Jr.* (12–15). HarperCollins.

Kiperman, S., DeLong, G., Varjas, K., & Meyers, J. (2021). Providing inclusive strategies for practitioners and researchers working with gender and sexually diverse youth without parental/guardian consent. In M. C. Lytle & R. A. Sprott (Eds.), *Supporting gender identity and sexual orientation diversity in K–12 schools* (pp. 97–120). American Psychological Association.

Kiperman, S., Lorig, C. E., & Cullen, E. (2025). A call to disrupt hetero- and cisnormativity (HetCisNorms) in school psychology with guidance from the Adapting Strategies to Promote Implementation Reach and Equity (ASPIRE) framework. *School Psychology, 40,* 622–635.

Kirby, G., Dolan, A., & Howard, H. (2022, April 29). An extraordinary, dangerous vandals' charter: MPs slam verdict in Colston statue destroying case and government vows to keep prosecuting suspects as mob of BLM activists who tore it down spout woke platitudes on the court steps after walking free. *Daily Mail.* https://www.dailymail.co.uk/news/article-10373391/Colston-statue-trial-MPs-slam-verdict-mob-BLM-activists-spout-woke-platitudes-court-steps.html

Kirk, C. [@charliekirk11]. (2018, October 1). *Liberals have: Feelings and emotion conservatives have: Logic, history, perspective, science, reason, facts, math, principles, ethics, values, dialogue & common* [Post]. Twitter. https://twitter.com/charliekirk11/status/1046581846174306309

Kirkland, J. (2025, March 3). 'The basis of eugenics': Elon Musk and the menacing return of the R-word. *The Guardian.* https://www.theguardian.com/world/2025/mar/03/r-word-right-wing-rise

Knopf, O. (1935). Preliminary report on personality studies in thirty migraine patients. *Journal of Nervous and Mental Disease, 82,* 270–285.

Koch, S. (1981). The nature and limits of psychological knowledge: Lessons of a century qua 'science'. *American Psychologist, 36,* 257–269.

Kolodinsky, R. W., Madden, T. M., Zisk, D. S., & Henkel, E. T. (2010). Attitudes about corporate social responsibility: Business student predictors. *Journal of Business Ethics, 91,* 167–181.

Kotik-Friedgut, B., & Ardila, A. (2020). A.R. Luria's cultural neuropsychology in the 21st century. *Culture & Psychology, 26,* 274–286.

Kraepelin, E. (1908). Zur Entartungsfrage. *Zentralblatt Für Nervenheilkunde Und Psychiatrie, 31,* 745–751.

Ku, L., Dittmar, H., & Banerjee, R. (2014). To have or to learn? The effects of materialism on British and Chinese children's learning. *Journal of Personality and Social Psychology, 106,* 803–821.

Kwapong, Y. A., Boakye, E., Khan, S. S., Honigberg, M. C., Martin, S. S., Oyeka, C. P., Hays, A. G., Natarajan, P., Mamas, M. A., Blumenthal, R. S., Blaha, M. J., & Sharma, G. (2023). Association of depression and poor mental health with cardiovascular disease and suboptimal cardiovascular health among young adults in the United States. *Journal of the American Heart Association, 12,* e028332.

Laird, A. R. (2021). Large, open datasets for human connectomics research: Considerations for reproducible and responsible data use. *NeuroImage, 244,* 118579.

Langbert, M., Quain, A. J., & Klein, D. B. (2016). Faculty voter registration in economics, history, journalism, law, and psychology. *Economics Journal Watch, 13,* 422–451.

Langbert, M., & Stevens, S. (2020). Partisan registration and contributions of faculty in flagship colleges. *Studies in Higher Education, 47,* 1750.

Langworthy, O. R. (1948). Relation of personality problems to onset and progress of multiple sclerosis. *Archives of Neurology & Psychiatry, 59,* 13–28.

Lanz, T. V., Brewer, R. C., Ho, P. P., Moon, J. S., Jude, K. M., Fernandez, D., Fernandes, R. A., Gomez, A. M., Nadj, G. S., Bartley, C. M., Schubert, R. D., Hawes, I. A., Vazquez, S. E., Iyer, M., Zuchero, J. B., Teegen, B., Dunn, J. E., Lock, C. B., Kipp, L. B., ... Robinson, W. H., et al. (2022). Clonally expanded B cells in multiple sclerosis bind EBV EBNA1 and GlialCAM. *Nature, 603,* 321–327.

LaRocca, N. G. (1984). Psychosocial factors in multiple sclerosis and the role of stress. *Annals of the New York Academy of Sciences, 436,* 435–442.

Larzelere, R. E., Reitman, D., Ortiz, C., & Cox Jr., R. B. (2023). Parental punishment: Don't throw out the baby with the bathwater. In C. L. Frisby, R. E. Redding, W. T. O'Donohue, & S. O. Lilienfeld (Eds.), *Ideological and political bias in psychology* (pp. 561–583). Springer.

Lavery, G. (2021, October 17). The UK media has seriously bungled the Kathleen Stock story. *Grace Lavery*. https://www.gracelavery.org/uk-media-biased-stock-sussex/

Le Bon, G. (1879). Recherches anatomiques et mathématiques sur les variations de volume du cerveau et sur leurs relations avec l'intelligence. *Revue d'Anthropologie, 8*(2), 27–104.

Le Duc, F. (2022, July 17). Royal honour for sussex professor forced out in trans row. *Brighton and Hove News*. https://www.brightonandhovenews.org/2022/07/17/royal-honour-for-sussex-professor-forced-out-in-trans-row/

Lee, B. Y. (2019, February 27). Selma Blair: Why her multiple sclerosis diagnosis was delayed. *Forbes*. https://www.forbes.com/sites/brucelee/2019/02/27/selma-blair-why-her-multiple-sclerosis-diagnosis-was-delayed/

Letzter, R. (2016, September 22). Scientists are furious after a famous psychologist accused her peers of 'methodological terrorism'. *Business Insider*. http://uk.businessinsider.com/susan-fiske-methodological-terrorism-2016-9

Lidz, T., & Whitehorn, J. C. (1950). Life situations, emotions, and Graves' disease. *Psychosomatic Medicine, 12*, 184–186.

Lightfoot, L. (2018, July 17). Universities outsource mental health services despite soaring demand. *The Guardian*. https://www.theguardian.com/education/2018/jul/17/universities-outsource-mental-health-services-despite-soaring-demand

Lin, L., Stamm, K., & Christidis, P. (2018). How diverse is the psychology workforce? News from APA's Center for Workforce Studies. *Monitor on Psychology, 49*, 19.

Lindberg, S. M., Hyde, J. S., Petersen, J., & Linn, M. C. (2010). New trends in gender and mathematics performance: A meta-analysis. *Psychological Bulletin, 136*, 1123–1135.

Lindsay, J. (2020, October 25). There's no such thing as 'systematic racism'. *New Discourses*. https://newdiscourses.com/2020/10/theres-no-such-thing-as-systematic-racism/

Liu, C. (2021). *Virtue hoarders: The case against the professional managerial class*. University of Minnesota Press.

Longino, H. E. (2002). *The fate of knowledge*. Princeton University Press.

Lorde, A. (1984). *Sister outsider: Essays and speeches*. Crossing Press.

Löve, J., Bertilsson, M., Martinsson, J., Wängnerud, L., & Hensing, G. (2019). Political ideology and stigmatizing attitudes toward depression: The Swedish case. *International Journal of Health Policy and Management*, *8*, 365–374.

Lovett, B. J. (2006). The new history of psychology: A review and critique. *History of Psychology*, *9*, 17–37.

Lukianoff, G. (2014). *Unlearning liberty: Campus censorship and the end of American debate*. Encounter Books.

Lukianoff, G., & Haidt, J. (2018). *The coddling of the American mind: How good intentions and bad ideas are setting up a generation for failure*. Allen Lane.

Lund, C., Breen, A., Flisher, A. J., Kakuma, R., Corrigall, J., Joska, J. A., Swartz, L., & Patel, V. (2010). Poverty and common mental disorders in low and middle income countries: A systematic review. *Social Science and Medicine*, *71*, 517–528.

Lyons, C. (2020, June 15). Petition seeks removal of MSU VP of research over controversial comments, research. *Lansing State Journal*. https://eu.lansing statejournal.com/story/news/2020/06/15/michigan-state-msu-stephen-hsu-research-removal-petition-graduate-employees-union/5345120002/

Macaulay, T. (1835). Minute by the Hon'ble T. B. Macaulay, dated the 2nd February 1835. In H. Sharp (Ed.), *Selections from educational records, part I* (1781–1839). Superintendent, Government Printing. [1919].

Mackenzie, D. (2004, November 3). US in U-turn over Gulf war syndrome. *New Scientist*. https://www.newscientist.com/article/dn6609-us-in-u-turn-over-gulf-war-syndrome/

MacLeod, C., & Campbell, L. (1992). Memory accessibility and probability judgments: An experimental evaluation of the availability heuristic. *Journal of Personality and Social Psychology*, *63*, 890–902.

Maher, M. (1890). *Psychology*. Longmans, Green & Co.

Major, B., Appelbaum, M., Beckman, L., Dutton, M. A., Russo, N. F., & West, C. (2009). Abortion and mental health: Evaluating the evidence. *American Psychologist*, *64*, 863–890.

Makel, M. C., Plucker, J. A., & Hegarty, B. (2012). Replications in psychology research: How often do they really occur? *Perspectives on Psychological Science*, *7*, 537–542.

Malik, K. (2024, November 10). Cosplaying social justice is the new elitist way of elbowing out the working class. *The Guardian*. https://www.theguardian.com/commentisfree/2024/nov/10/cosplaying-social-justice-is-the-new-elitist-way-of-elbowing-out-the-working-class

Malik, K. (2025, April 27). Trans rights: 'MPs are outsourcing their morals'. *The Observer.* https://observer.co.uk/news/opinion-and-ideas/article/kenan-malik-column-trans-rights-supreme-court

Malmusi, D., Palència, L., Ikram, U. Z., Kunst, A. E., & Borrell, C. (2017). Inequalities by immigrant status in depressive symptoms in Europe: The role of integration policy regimes. *Social Psychiatry and Psychiatric Epidemiology, 52,* 391–398.

Malone, C. (2024, July 19). These eco-activists are just narcissists: The judge has given them what they deserve. *Daily Express.* https://www.express.co.uk/comment/columnists/carole-malone/1925493/eco-activists-jailed-m25-protest-reaction

Mance, H. (2016, June 3). Britain has had enough of experts, says Gove. *Financial Times.* https://www.ft.com/content/3be49734-29cb-11e6-83e4-abc22d5d108c

Manstead, A. S. R. (2018). The psychology of social class: How socioeconomic status impacts thoughts, feelings, and behaviour. *British Journal of Social Psychology, 57,* 267–291.

Marks, D. F. (2019). The Hans Eysenck affair: Time to correct the scientific record. *Journal of Health Psychology., 24,* 409–420.

Marková, I. (2012). 'Americanization' of European social psychology. *History of the Human Sciences, 25,* 108–116.

Marsh, S. (2021, October 29). ME exercise therapy guidance scrapped by health watchdog Nice. *The Guardian.* https://www.theguardian.com/society/2021/oct/29/health-watchdog-nice-publishes-delayed-me-guidance

Martin, J. D., Song, S. W., Rote, W. R., Bakour, C., Rance, L. T., Scacco, J. M., & Marcus, S. (2025). *The life in media survey: A baseline study of digital media use and well-being among 11- to 13-year-olds.* Tampa: University of South Florida.

Maryanski, A. (2010). WEIRD societies may be more compatible with human nature. *Behavioral and Brain Sciences, 33,* 103–104.

Mazzei, P. (2023, February 14). DeSantis's latest target: A small college of 'free thinkers'. *New York Times.* https://www.nytimes.com/2023/02/14/us/ron-desantis-new-college-florida.html

McClintock, C. G., Spaulding, C. B., & Turner, H. A. (1965). Political orientations of academically affiliated psychologists. *American Psychologist, 20,* 211–221.

McFarling, U. L. (2021, September 23). 'Health equity tourists': How white scholars are colonizing research on health disparities. *STAT*. https://www.statnews.com/2021/09/23/health-equity-tourists-white-scholars-colonizing-health-disparities-research/

McNamara, M., Baker, K., Connelly, K., Janssen, A., Olson-Kennedy, J., Pang, K. C., Scheim, A., Turban, J., & Alstott, A. (2024). *An evidence-based critique of 'The Cass Review' on gender-affirming care for adolescent gender dysphoria*. Yale Law School.

McNulty, J. K., & Fincham, F. D. (2012). Beyond positive psychology? Toward a contextual view of psychological processes and well-being. *American Psychologist, 67*, 101–110.

McVeigh, T. (2011, August 6). Research linking autism to internet use is criticised. *The Guardian*. https://www.theguardian.com/society/2011/aug/06/research-autism-internet-susan-greenfield

Melchior, J. K. (2020, June 25). A Twitter mob takes down administrator at Michigan State. *Wall Street Journal*. https://www.wsj.com/articles/a-twitter-mob-takes-down-an-administrator-at-michigan-state-11593106102

Merlan, A. (2025, February 4). Elon Musk's midnight takeover. *Mother Jones*. https://www.motherjones.com/politics/2025/02/elon-musk-takeover-doge-gsa-opm-secrecy-dei/

Mervis, J. (2025, February 5). Trump orders cause chaos at science agencies. *Science*. https://www.science.org/content/article/trump-orders-cause-chaos-science-agencies

Metzl, J. M. (2010). *The protest psychosis: How schizophrenia became a Black disease*. Beacon Press.

Mezaki, T., & Namekawa, K. (2019). Jacqueline du Pré and her alternative diagnosis. *Medical Hypotheses, 133*, 109401.

Miller, J. (2012, November 19). Rubio on Earth's age: 'I'm not a scientist, man'. *CBS News*. https://www.cbsnews.com/news/rubio-on-earths-age-im-not-a-scientist-man/

Mills, C. (2007). White ignorance. In S. Sullivan & N. Tuana (Eds.), *Race and epistemologies of ignorance* (pp. 13–37). State University of New York Press.

Mitchell, K. J. (2018). *Innate: How the wiring of our brains shapes who we are*. Princeton University Press.

Moghaddam, F. M. (1987). Psychology in the three worlds: As reflected by the crisis in social psychology and the move toward indigenous third-world psychology. *American Psychologist, 42*, 912–920.

Moghaddam, F. M. (2023). *How psychologists failed: We neglected the poor and minorities, favored the rich and privileged, and got science wrong.* Cambridge University Press.

Monbiot, G. (2021a, January 21). We're about to see a wave of long Covid. When will ministers take it seriously? *The Guardian.* https://www.theguardian.com/commentisfree/2021/jan/21/were-about-to-see-a-wave-of-long-covid-when-will-ministers-take-it-seriously

Monbiot, G. (2021b, April 14). Apparently just by talking about it, I'm super-spreading long Covid. *The Guardian.* https://www.theguardian.com/commentisfree/2021/apr/14/super-spreading-long-covid-professor-press-coverage

Monbiot, G., & Hutchison, P. (2024). *The invisible doctrine: The secret history of neoliberalism (& how it came to control your life).* Allen Lane.

Moran, G. (1985). The origins and development of boycotting. *Journal of the Galway Archaeological and Historical Society, 40,* 49–64.

Morgan, E. (2016, December 15). Piers Morgan is wrong about Lady Gaga. Here are the facts about PTSD. *The Guardian.* https://www.theguardian.com/commentisfree/2016/dec/15/piers-morgan-ptsd-lady-gaga-post-traumatic-stress-disorder-soldiers

Morris, S. (2024, August 10). 'I've been tired since I was 13': ME patients hope harrowing inquest will change perceptions. *The Guardian.* https://www.theguardian.com/society/article/2024/aug/10/me-patients-maeve-boothby-o-neill-inquest

Moscovici, S. (1981). *The age of the crowd: A historical treatise on mass psychology.* Cambridge University Press.

Müller, J. -W. (2025, April 18). Harvard shows resistance is possible. But universities must join forces. *The Guardian.* https://www.theguardian.com/commentisfree/2025/apr/18/harvard-resistance-possible-universities

Munzel, M. (2002). Multiple sclerosis: The psychosomatic consequence of unsuccessful bonding. In C. D. Ventling (Ed.), *Body psychotherapy in progressive and clinical disorde* (pp. 35–48). Karger.

Muradian, R., & Gomez-Baggethun, E. (2021). Beyond ecosystem services and nature's contributions: Is it time to leave utilitarian environmentalism behind? *Ecological Economics, 185,* 107038.

Myers, P. Z. (2023, May 23). The insufferable pettiness of Anna Krylov. *Pharyngula.* https://freethoughtblogs.com/pharyngula/2023/05/23/the-insufferable-pettiness-of-anna-krylov/

Nadal, K. L. (2017). 'Let's get in formation': On becoming a psychologist–activist in the 21st century. *American Psychologist, 72,* 935–946.

Naím, M. (2022). *The revenge of power: How autocrats are reinventing politics for the twenty-first century.* St Martin's Griffin.

Ng, E. (2022). *Cancel culture: A critical analysis.* Palgrave Macmillan.

NICE (2021, October 18). Myalgic encephalomyelitis (or encephalopathy)/ chronic fatigue syndrome: Diagnosis and management—Roundtable discussion: Minutes. *National Institute for Health and Care Excellence.* https:// www.nice.org.uk/guidance/ng206/documents/minutes-31

Noone, C., Southgate, A., Ashman, A., Quinn, É., Comer, D., Shrewsbury, D., Ashley, F., Hartland, J., Paschedag, J., Gilmore, J., Kennedy, N., Woolley, T. E., Heath, R., Goulding, R., Simpson, V., Kiely, E., Coll, S., White, M., Grijseels, D. M., Ouafik, M., . . . McLamore, Q., et al. (2025). Critically appraising the Cass report: Methodological flaws and unsupported claims. *BMC Medical Research Methodology, 25*, 128.

Norris, J. I., Lambert, N. M., Nathan DeWall, C., & Fincham, F. D. (2012). Can't buy me love? Anxious attachment and materialistic values. *Personality and Individual Differences, 53*, 666–669.

Novak, B. (2021, April 27). Hungary transfers 11 universities to foundations led by Orban allies. *New York Times.* https://www.nytimes.com/2021/04/ 27/world/europe/hungary-universities-orban.html

Nović, S. (2025, March 27). The US right is coming for disabled people. Here's why that threatens everyone. *The Guardian.* https://www.theguar dian.com/global/2025/mar/27/us-disability-rights-trump

O'Connor, K. (2013). *Pineapple: A global history.* Reaktion Books.

O'Donohue, W. T. (2023). Prejudice and the quality of the science of contemporary social justice efforts in psychology. In C. L. Frisby, R. E. Redding, W. T. O'Donohue, & S. O. Lilienfeld (Eds.), *Ideological and political bias in psychology* (pp. 173–200). Springer.

O'Leary, D., & Geraghty, K. (2021). Ethical psychotherapeutic management of patients with medically unexplained symptoms: The risk of misdiagnosis and harm. In M. Trachsel, J. Gaab, N. Biller-Andorno, Ş. Tekin, & J. Z. Sadler (Eds.), *The Oxford handbook of psychotherapy ethics* (pp. 789–802). Oxford University Press.

O'Loughlin, A. (2022, June 15). Cork University Hospital settles case for €6m over delay in diagnosing brain tumour in teenager. *Irish Examiner.* https:// www.irishexaminer.com/news/courtandcrime/arid-40896041.html

Oakeshott, I. (2024, October 10). These art vandals are just self-indulgent idiots. *Daily Telegraph.* https://www.telegraph.co.uk/news/2024/10/10/art-vandals-self-indulgent-idiots-protests-national-gallary/

Odgers, C. L. (2024). The great rewiring, unplugged. *Nature*, *628*, 29–30.

Office for National Statistics (UK) (2023, March 28). Ethnic group. *Census 2021*. https://www.ons.gov.uk/datasets/TS021/editions/2021/versions/3/filter-outputs/2c225a7b-0b5a-4a56-825e-2d6df1c6be93

Office for Veterans Affairs (2023, October 3). *Expert meeting on the health conditions of Gulf War Veterans*. https://www.gov.uk/government/news/expert-meeting-on-the-health-conditions-of-gulf-war-veterans

Oksanen, A., Celuch, M., Latikka, R., Oksa, R., & Savela, N. (2022). Hate and harassment in academia: The rising concern of the online environment. *Higher Education*, *84*, 541–567.

Olszewski, T. M., & Varrasse, J. F. (2005). The neurobiology of PTSD: Implications for nurses. *Journal of Psychosocial Nursing and Mental Health Services*, *43*, 40–47.

Open Science Collaboration. (2015). Estimating the reproducibility of psychological science. *Science*, *349*, aac4716.

Ornstein, R. (1988). *Psychology: The study of human experience* (2nd ed.). Harcourt Brace Jovanovich.

Osborne, S. (2023, October 14). Boris Johnson asked if government 'believes in long COVID', coronavirus inquiry hears. *Sky News*. https://news.sky.com/story/boris-johnson-asked-if-government-believes-in-long-covid-coronavirus-inquiry-hears-12984294

Palmer, C. (2021). The top 10 journal articles of 2020. *Monitor on Psychology*, *52*, 24.

Pankratz, A. (2024, June 19). Just stop oil's deranged Stonehenge attack goes beyond mere idiocy. *National Post*. https://nationalpost.com/opinion/adam-pankratz-just-stop-oils-deranged-stonehenge-attack-goes-beyond-mere-idiocy

Panofsky, A., Dasgupta, K., & Iturriaga, N. (2021). How White nationalists mobilize genetics: From genetic ancestry and human biodiversity to counterscience and metapolitics. *American Journal of Physical Anthropology*, *175*, 387–398.

Parker, I. (2009). Critical psychology and revolutionary Marxism. *Theory & Psychology*, *19*, 71–92.

Parsons, V. (2021, January 27). MPs urged by anti-trans 'women's rights' group to eliminate 'transgenderism' and scrap Gender Recognition Act. *PinkNews*. https://www.thepinknews.com/2021/01/27/womens-human-rights-campaign-gender-recognition-act-inquiry-trans-transphobia/

Patel, N., Alcock, K., Alexander, L., Baah, J., Butler, C., Danquah, A., Gibbs, A., Goodbody, L., Joseph-Loewenthal, W., Muhxinga, Z., Ong, L., Peart, A., Rennalls, S., Tong, K., & Wood, N. (2019). Racism is not entertainment. *Psychologists for Social Change*, https://www.psychchange.org/racism-is-not-entertainment.html.

Pearce, J. (1977). Migraine: A psychosomatic disorder. *Headache: The Journal of Head and Face Pain, 17*, 125–128.

Pearson, K. (1901). *National life: From the standpoint of science.* Adam and Charles Black.

Pellicane, M. J., & Ciesla, J. A. (2022). Associations between minority stress, depression, and suicidal ideation and attempts in transgender and gender diverse (TGD) individuals: Systematic review and meta-analysis. *Clinical Psychology Review., 91*, 102113.

Pelosi, A. J. (2019). Personality and fatal diseases: Revisiting a scientific scandal. *Journal of Health Psychology., 24*, 421–439.

Pengelly, M. (2024a, April 8). Trump bemoans lack of immigrants from majority-white countries to the US. *The Guardian.* https://www.theguardian.com/us-news/2024/apr/08/trump-immigration-north-europe

Pengelly, M. (2024b, August 6). JD Vance pleads sarcasm in latest effort to clean up 'childless cat ladies' remark. *The Guardian.* https://www.theguardian.com/us-news/article/2024/aug/06/jd-vance-childless-cat-ladies-tim-walz

Pepin-Neff, C., & Cohen, A. (2021). President Trump's transgender moral panic. *Policy Studies, 42*, 646–661.

Peplau, L. A., & Fingerhut, A. W. (2007). The close relationships of lesbians and gay men. *Annual Review of Psychology, 58*, 405–424.

Pérez-Álvarez, M. (2013). Positive psychology and its friends: Revealed. *Papeles del Psicólogo, 34*, 208–226.

Perlin, R. (2024). *Language city: The fight to preserve endangered mother tongues.* Grove Press.

Perry, S. (2025, April 9). Wes Streeting 'sorry' for the 'fear' caused by puberty blocker ban — but he stands by it. *Pink News.* https://www.thepinknews.com/2025/04/09/wes-streeting-puberty-blockers-ban-trans-sorry/

Peterson, J. (2023, December 14). Pro-Hamas protesters are sanctimonious psychopaths. *Daily Telegraph.* https://www.telegraph.co.uk/news/2023/12/14/jordan-peterson-pro-hamas-protesters-sanctimonious-psychopa/

Petroski, H. (1990). *The pencil: A history of design and circumstance.* Knopf.

Pickren, W. E. (2000). A whisper of salvation: American psychologists and religion in the popular press, 1884-1908. *American Psychologist, 55*, 1022–1024.

Pickren, W. E. (2009). Indigenization and the history of psychology. *Psychological Studies, 54*, 87–95.

Pieters, R. (2013). Bidirectional dynamics of materialism and loneliness: Not just a vicious cycle. *Journal of Consumer Research, 40*, 615–631.

Pietraszewski, D., & Wertz, A. E. (2022). Why evolutionary psychology should abandon modularity. *Perspectives on Psychological Science, 17*, 465–490.

Pilgrim, D. (2024, August 30). Trans capture in the BPS in its social and historical context. *BPS Watch.* https://bpswatch.com/2024/08/30/trans-capture-in-the-bps-in-its-social-and-historical-context/

Pinker, S. (2006, June 26). Groups and genes. *The New Republic.* https://newrepublic.com/article/77727/groups-and-genes

Pinker, S. (2018). *Enlightenment now: The case for reason, science, humanism and progress.* Penguin.

Pinker, S.[@sapinker] (2025, April 13). *Sad to agree with @Cathyyoung63 and @Mattjj89 that the Free Press (for which i had high hopes) has 'descended into* [Post]. Twitter. https://twitter.com/sapinker/status/1911548393970622727

Place, N. (2021, November 9). Kathleen Stock: Professor accused of transphobia takes job at 'anti-woke' University of Austin. *The Independent.* https://www.independent.co.uk/news/world/americas/kathleen-stock-transphobia-university-austin-b1954486.html

Porter, R. (1987). *A social history of madness: The world through the eyes of the insane.* Weidenfeld & Nicolson.

Prentice, D. A. (2012). Liberal norms and their discontents. *Perspectives on Psychological Science, 7*, 516–518.

Prilleltensky, I. (1994). *The morals and politics of psychology: Psychological discourse and the status quo.* State University of New York Press.

Pring, J. (2022, September 29). Labour conference: Anger at Starmer's 'divisive' pledge to back 'working people'. *Disability News Service.* https://www.disabilitynewsservice.com/labour-conference-anger-at-starmers-divisive-pledge-to-back-working-people/

Pringle, P. (1996, October 31). Eysenck took £800,000 tobacco funds. *The Independent.* https://www.independent.co.uk/news/eysenck-took-pounds-800-000-tobacco-funds-1361007.html

Prugh, D. G. (1951). A preliminary report on the role of emotional factors in idiopathic celiac disease. *Psychosomatic Medicine, 13*, 220–241.

Quinn, B. (2024, April 2). Rough sleepers 'should not be arrested just if they smell', says UK minister. *The Guardian.* https://www.theguardian.com/society/2024/apr/02/rough-sleepers-should-not-be-arrested-just-if-they-smell-says-uk-minister

Quinn, R. (2025, February 27). Trump is targeting DEI in higher ed. But what does he mean? *Inside Higher Ed.* https://www.insidehighered.com/news/diversity/2025/02/27/trump-targeting-dei-higher-ed-what-does-he-mean

Ramirez-Simon, D., Yerushalmy, J., Helmore, E., & Salam, E. (2024, May 1). Dozens arrested at Columbia University as New York police disperse Gaza protest. *The Guardian.* https://www.theguardian.com/us-news/2024/apr/30/new-york-police-columbia-university-student-protests

Rea, M., & Tabor, D. (2022, April 25). Health state life expectancies by national deprivation deciles, England: 2018 to 2020. *Office for National Statistics: Census 2021.* https://www.ons.gov.uk/peoplepopulationandcommunity/healthandsocialcare/healthinequalities/bulletins/healthstatelifeexpectanciesbyindexofmultipledeprivationimd/2018to2020

Reay, D. (2018). Working class educational transitions to university: The limits of success. *European Journal of Education, 53,* 538–540.

Redden, E. (2019, May 6). In Brazil, a hostility to academe. *Inside Higher Ed.* https://www.insidehighered.com/news/2019/05/06/far-right-government-brazil-slashes-university-funding-threatens-cuts-philosophy-and

Redding, R. E. (2001). Sociopolitical diversity in psychology: The case for pluralism. *American Psychologist, 56,* 205–215.

Redding, R. E. (2023). Psychologists' politics. In C. L. Frisby, R. E. Redding, W. T. O'Donohue, & S. O. Lilienfeld (Eds.), *Ideological and political bias in psychology* (pp. 79–95). Springer.

Reed, E. (2025, March 22). Military will now ask soldiers if they have 'symptoms' of being trans. *Erin in the Morning.* https://www.erininthemorning.com/p/military-will-now-ask-soldiers-if

Reuveny, R. (2008). Ecomigration and violent conflict: Case studies and public policy implications. *Human Ecology, 36,* 1–13.

Rhodes, C. (2022, February 3). 'Woke' business schools? The reality is quite the opposite. *Times Higher Education.* https://www.timeshighereducation.com/blog/woke-business-schools-reality-quite-opposite

Richards, G., & Stenner, P. (2023). *Putting psychology in its place: Critical historical perspectives* (4th ed.). Routledge.

Richins, M. L. (2004). The positive and negative consequences of materialism: What are they and when do they occur? *Advances in Consumer Research, 31,* 232–235.

Ridley, M., Rao, G., Schilbach, F., & Patel, V. (2020). Poverty, depression, and anxiety: Causal evidence and mechanisms. *Science, 370,* eaay0214.

Rigney, D. (1991). Three kinds of anti-intellectualism: Rethinking Hofstader. *Sociological Inquiry, 61,* 434–451.

Riley, S., Frith, H., Archer, L., & Veseley, L. (2006). Institutional sexism in academia. *The Psychologist, 19,* 94–97.

Rimer, S., & Healy, P. D. (2005, February 18). Furor lingers as Harvard chief gives details of talk on women. *New York Times.* https://www.nytimes.com/2005/02/18/education/furor-lingers-as-harvard-chief-gives-details-of-talk-on-women.html

Roberts, S. O., Bareket-Shavit, C., Dollins, F. A., Goldie, P. D., & Mortenson, E. (2020). Racial inequality in psychological research: Trends of the past and recommendations for the future. *Perspectives on Psychological Science, 15,* 1295–1309.

Robertson, D. (2021, March 21). How 'owning the libs' became the GOP's core belief. *Politico.* https://www.politico.com/news/magazine/2021/03/21/owning-the-libs-history-trump-politics-pop-culture-477203

Robinson, A. (2016). The Marconi connection. *New Scientist, 3086,* 42–43.

Robinson, D. N. (2013). Historiography in psychology: A note on ignorance. *Theory & Psychology, 23,* 819–828.

Rogers, R. A. (1993). Pleasure, power and consent: The interplay of race and gender in *New Jack City. Women's Studies in Communication, 16,* 62–85.

Rogers, S. (2025, January 15). Top ten most popular courses in the UK. *Complete University Guide.* https://www.thecompleteuniversityguide.co.uk/student-advice/what-to-study/top-ten-most-popular-courses-in-the-uk

Rondinella, S., & Silipo, D. B. (2023). Income dissatisfaction and migraine headache: Evidence from a nationwide population-based survey. *Health Psychology and Behavioral Medicine, 11,* 2266214.

Rose, D., & Beresford, P. (2024). PPI in psychiatry and the problem of knowledge. *BMC Psychiatry, 24,* 52.

Rosenthal, A. (1988, August 16). Dukakis asserts that his 'L word' is 'leadership'. *New York Times.* https://www.nytimes.com/1988/08/16/us/dukakis-asserts-that-his-l-word-is-leadership.html

Roser, M. [@MaxCRoser]. (2023, March 29). Terrible to see the many reports about rising suicide rates among young people in the US. I wanted to see

[Image attached][Post]. *Twitter.* https://twitter.com/MaxCRoser/status/ 1641146860281659406

Ross, L. (2015). What kinds of conservatives does social psychology lack, and why? *Behavioral and Brain Sciences, 38*, e156.

Rutherford, A. (2022). *Control: The dark history and troubling present of Eugenics.* Weidenfeld & Nicolson.

Sagan, C. (1997). *The demon-haunted world: Science as a candle in the dark.* Ballantine.

Salo, S. K., Marceaux, J. C., McCoy, K. J. M., & Hilsabeck, R. C. (2022). Removing the noose item from the Boston Naming Test: A step toward antiracist neuropsychological assessment. *Clinical Neuropsychologist, 36*, 311–326.

Sánchez-Román, S., Téllez-Zenteno, J. F., Zermeño-Phols, F., García-Ramos, G., Velázquez, A., Derry, P., Hernández, M., Resendiz, A., & Guevara-López, U. M. (2007). Personality in patients with migraine evaluated with the 'Temperament and Character Inventory'. *Journal of Headache and Pain, 8*, 94–104.

Sareen, J., Afifi, T. O., McMillan, K. A., & Asmundson, G. J. G. (2011). Relationship between household income and mental disorders: Findings from a population-based longitudinal study. *Archives of General Psychiatry, 68*, 419–427.

Sarkar, A. (2025). *Minority rule: Adventures in the Culture War.* Bloomsbury.

Savitz, J. B., Rauch, S. L., & Drevets, W. C. (2013). Clinical applications of brain imaging for the diagnosis of mood disorders. *Molecular Psychiatry, 18*, 528–539.

Scamvougeras, A., & Howard, A. (2020). Somatic symptom disorder, medically unexplained symptoms, somatoform disorders, functional neurological disorder: How DSM-5 got it wrong. *Canadian Journal of Psychiatry/Revue Canadienne de Psychiatrie, 65*, 301–305.

Schmidt, C. G. (1984). The group-fantasy origins of AIDS. *Journal of Psychohistory, 12*, 37–78.

Schneider, A., & Ingram, H. (2005). *Deserving and entitled: Social constructions and public policy.* State University of New York Press.

Schofield, P., Thygesen, M., Das-Munshi, J., Becares, L., Cantor-Graae, E., Pedersen, C., & Agerbo, E. (2017). Ethnic density, urbanicity and psychosis risk for migrant groups: A population cohort study. *Schizophrenia Research., 190*, 82–87.

Scott, K. M., Lim, C., Al-Hamzawi, A., Alonso, J., Bruffaerts, R., Caldas-de-almeida, J. M., Florescu, S., de Girolamo, G., Hu, C., de Jonge, P., Kawakami,

N., Medina-Mora, M. E., Moskalewicz, J., Navarro-Mateu, F., O'Neill, S., Piazza, M., Posada-Villa, J., Torres, Y., & Kessler, R. C. (2016). Association of mental disorders with subsequent chronic physical conditions: World mental health surveys from 17 countries. *JAMA Psychiatry.*, *73*, 150–158.

Seibt, B., Waldzus, S., Schubert, T. W., & Brito, R. (2015). Conservatism is not the missing viewpoint for true diversity. *Behavioral and Brain Sciences*, *38*, e157.

Seligman, M. E. P. (2019). Positive psychology: A personal history. *Annual Review of Clinical Psychology*, *15*, 1–23.

Shakespeare, T., Watson, N., & Abu Alghaib, O. (2017). Blaming the victim, all over again: Waddell and Aylward's biopsychosocial (BPS) model of disability. *Critical Social Policy*, *37*, 22–41.

Sham Ku, K. (2020). A culture of silence and denial. *The Psychologist*, *33*, 2.

Shapiro, F. (2021). *The New Yale book of quotations*. Yale University Press.

Sheldon, K. M., & Kasser, T. (1995). Coherence and congruence: Two aspects of personality integration. *Journal of Personality and Social Psychology*, *68*, 531–543.

Shonkoff, J. P., & Garner, A. S. (2012). Committee on psychosocial aspects of child and family health, committee on early childhood, adoption, and dependent care, & section on developmental and behavioral pediatrics the lifelong effects of early childhood adversity and toxic stress. *Pediatrics*, *129*, e232–e246.

Shulman, S. (2006). *Undermining science: Suppression and distortion in the bush administration*. University of California Press.

Siegel, E. (2025). How glorifying ignorance leads to science illiteracy. *Big Think*. https://bigthink.com/starts-with-a-bang/glorifying-ignorance-scientific-illiteracy/

Silander, N. C., Geczy Jr., B., Marks, O., & Mather, R. D. (2020). Implications of ideological bias in social psychology on clinical practice. *Clinical Psychology: Science and Practice*, *27*, e12312.

Singh, S. P., Greenwood, N., White, S., & Churchill, R. (2007). Ethnicity and the Mental Health Act 1983: Systematic review. *British Journal of Psychiatry*, *191*, 99–105.

Smaldino, P. E., & McElreath, R. (2016). The natural selection of bad science. *Royal Society Open Science*, *3*, 160384.

Smith, J. (2024, August 9). Treatment changes urged after ME patient's death. *BBC News*. https://www.bbc.com/news/articles/czrgmdv4z0go

Society for Personality and Social Psychology (2019, January). *SPSP diversity and climate survey: Final report.* https://spsp.org/sites/default/files/SPSP_Diversity_and_Climate_Survey_Final_Report_January_2019.pdf

Sokal, A., & Bricmont, J. (1998). *Fashionable nonsense: Postmodern intellectuals' abuse of science.* Picador.

Solberg, E. C., Diener, E., & Robinson, M. D. (2004). Why are materialists less satisfied? In T. Kasser & A. D. Kanner (Eds.), *Psychology and consumer culture: The struggle for a good life in a materialistic world* (pp. 29–48). American Psychological Association.

Southern Poverty Law Center (2023, December 12). Combating anti-LGBTQ+ pseudoscience. *Southern Poverty Law Center.* https://www.splcenter.org/resources/guides/captain/

Spivey, L. A., & Edwards-Leeper, L. (2019). Future directions in affirmative psychological interventions with transgender children and adolescents. *Journal of Clinical Child & Adolescent Psychology, 48,* 343–356.

Sridhar, M. (2024, December 28). Modern psychology and its colonial legacy. *Mad in America.* https://www.madinamerica.com/2024/12/modern-psychology-and-its-colonial-legacy/

Stanley, J. (2018, September 21). Hungary to ban gender studies degrees. *University Observer.* https://universityobserver.ie/hungary-to-ban-gender-studies-degrees/

Stapleton, G. (2017). Qualifying choice: Ethical reflection on the scope of prenatal screening. *Medicine, Health Care, and Philosophy, 20,* 195–205.

Steeg, S., Carr, M. J., Mok, P. L. H., Pedersen, C. B., Antonsen, S., Ashcroft, D. M., Kapur, N., Erlangsen, A., Nordentoft, M., & Webb, R. T. (2020). Temporal trends in incidence of hospital-treated self-harm among adolescents in Denmark: National register-based study. *Social Psychiatry & Psychiatric Epidemiology, 55,* 415–421.

Stephens, M. (1998). *The rise of the image and the fall of the word.* Oxford University Press.

Stevens, S. T., Jussim, L., & Honeycutt, N. (2020). Scholarship suppression: Theoretical perspectives and emerging trends. *Societies, 10,* 82.

Strand, M., Bäärnhielm, S., Fredlund, P., Brynedal, B., & Welch, E. (2023). Migration background, eating disorder symptoms and healthcare service utilisation: Findings from the Stockholm Public Health Cohort. *BJPsych Open, 9,* e205.

Strimpel, Z. (2024, May 4). The far-left mob has seized control of Britain's streetsand its borders. *Daily Telegraph*. https://www.telegraph.co.uk/news/2024/05/04/asylum-immigration-enforcement-protest-far-left-mob/

Stringer, C. (2023, January 15). Half of our universities peddle their woke agenda to students: League table reveals 'dark shadow' has fallen on elite institutions. *Daily Mail*. https://www.dailymail.co.uk/news/article-11638389/Half-universities-peddle-woke-agenda-students.html

Sue, S., & Zane, N. (2006). Ethnic minority populations have been neglected by evidence-based practices. In J. C. Norcross, L. E. Beutler, & R. F. Levant (Eds.), *Evidence-based practices in mental health: Debate and dialogue on the fundamental questions* (pp. 329–337). American Psychological Association.

Sugarman, J. (2015). Neoliberalism and psychological ethics. *Journal of Theoretical and Philosophical Psychology*, *35*, 103–116.

Sullivan, H. (2024, August 1). 'It was the same old show': Kamala Harris responds to Trump's attacks on her racial identity. *The Guardian*. https://www.theguardian.com/us-news/article/2024/aug/01/kamala-harris-texas-rally-donald-trump-racial-identity

Sutton, J. (2020). Editorial, September 2020. *The Psychologist*. https://www.bps.org.uk/psychologist/editorial-september-2020

Swain, F. (2011, July 27). Susan Greenfield: Living online is changing our brains. *New Scientist*. https://www.newscientist.com/article/mg21128236-400-susan-greenfield-living-online-is-changing-our-brains%20/

Sykes, M. A., Blanchard, E. B., Lackner, J., Keefer, L., & Krasner, S. (2003). Psychopathology in irritable bowel syndrome: Support for a psychophysiological model. *Journal of Behavioral Medicine*, *26*, 361–372.

Szabo, C. (2025). *Unreliable: Bias, fraud, and the reproducibility crisis in biomedical research*. Columbia University Press.

Tafreshi, D., Slaney, K. L., & Neufeld, S. D. (2016). Quantification in psychology: Critical analysis of an unreflective practice. *Journal of Theoretical and Philosophical Psychology*, *36*, 233–249.

Táíwò, O. O. (2022). *Elite capture: How the powerful took over identity politics (and everything else)*. Pluto Press.

Tattersall, A. (2019, April 8). Don't be a giraffe: How to avoid trolls on academic social media. *LSE Impact Blog*. https://blogs.lse.ac.uk/impactofsocialsciences/2019/04/08/dont-be-a-giraffe-how-to-avoid-trolls-on-academic-social-media/

Teghtsoonian, K. (2009). Depression and mental health in neoliberal times: A critical analysis of policy and discourse. *Social Science & Medicine*, *69*, 28–35.

Tensley, B. (2020, September 22). The dark subtext of Trump's 'good genes' compliment. *CNN Politics*. https://edition.cnn.com/2020/09/22/politics/donald-trump-genes-historical-context-eugenics/index.html

Teo, T. (2018). *Homo neoliberalus*: From personality to forms of subjectivity. *Theory & Psychology, 28*, 581–599.

Teo, T. (2022). What is a white epistemology in psychological science? A critical race-theoretical analysis. *Frontiers in Psychology, 13*, 861584.

Terman, L. M. (1916). *The measurement of intelligence: An explanation of and a complete guide for the use of the Stanford revision and extension of the Binet-Simon intelligence scale*. Houghton Mifflin.

Tetlock, P. E., & Mitchell, G. (2015). Why so few conservatives and should we care? *Society, 52*, 28–34.

Theodossopoulos, D. (2014). On de-pathologizing resistance. *History and Anthropology, 25*, 415–430.

Thibodeau, P. H., Fein, M. J., Goodbody, E. S., & Flusberg, S. J. (2015). The depression schema: How labels, features, and causal explanations affect lay conceptions of depression. *Frontiers in Psychology, 6*, 1728.

Thomason, A. (2023, January 18). Florida's state colleges say they'll ban promotion of critical race theory. *Chronicle of Higher Education*. https://www.chronicle.com/article/floridas-state-colleges-say-theyll-stop-promoting-critical-race-theory

Thompson, H. E. (2021). Climate 'psychopathology'. *European Psychologist, 26*, 195–203.

Thorbecke, C., Pereira, J., & Jones, E. (2019, February 26). Selma Blair reveals she cried with relief at MS diagnosis after being 'not taken seriously' by doctors. *ABC News*. https://abcnews.go.com/GMA/Culture/selma-blair-opens-tears-relief-ms-diagnosis/story?id=61310469

Thorndike, E. (1913). *Education psychology: Briefer course*. Teachers College.

Thorne, S. R., Hegarty, P., & Hepper, E. G. (2019). Equality in theory: From a heteronormative to an inclusive psychology of romantic love. *Theory & Psychology, 29*, 240–257.

Thrift, E., & Sugarman, J. (2019). What is social justice? Implications for psychology. *Journal of Theoretical and Philosophical Psychology, 39*, 1–17.

Tillett, E. (2018, July 24). Sessions slams liberal colleges for fostering 'sanctimonious, sensitive snowflakes'. *CBS News*. https://www.cbsnews.com/news/sessions-slams-liberal-colleges-for-fostering-sanctimonious-sensitive-snowflakes/

The Times. (2019, May 10). The Times view on the sacking of Noah Carl: Monoversities. *The Times.* https://www.thetimes.co.uk/article/the-times-view-on-the-sacking-of-noah-carl-monoversities-fhc7027qb

Touraine, G. A., & Draper, G. (1934). The migrainous patient: A constitutional study. *Journal of Nervous and Mental Disease, 80,* 1–23.

Treasure, R. A. R., Fowler, P. B. S., Millington, H. T., & Wise, P. H. (1987). Misdiagnosis of diabetic ketoacidosis as hyperventilation syndrome. *British Medical Journal, 294,* 630.

Tuller, D. (2022, June 4). Trial by error: Science Media Centre chief Fiona Fox compares ME/CFS patient advocates to Nazis. *Virology Blog.* https://virology.ws/2022/06/04/trial-by-error-science-media-centre-chief-compares-patient-advocates-to-nazis/

Tuller, D. (2024, August 6). Trial by error: Sean O'Neill's inquest statement. *Virology Blog.* https://virology.ws/2024/08/06/trial-by-error-sean-oneills-inquest-statement/

Turiel, E. (2020). Eugenics, prejudice, and psychological research. *Human Development, 64,* 103–107.

Tyson, C., & Oreskes, N. (2022). The American university, the politics of professors, and the narrative of 'liberal bias'. In J. Cruikshank & R. Abbinnett (Eds.), *The social production of knowledge in a neoliberal age: Debating the challenges facing higher education* (pp. 79–102). Rowman & Littlefield.

US Bureau of Labor Statistics (2024, January 26). Labor force statistics from the current population survey. *US Bureau of Labor Statistics.* https://www.bls.gov/cps/cpsaat11.htm

Van Peebles, M. (Director). (1991). *New Jack City* [Film]. Jackson/McHenry Company, Jacmac Films.

Van Voren, R. (2010). Political abuse of psychiatry: An historical overview. *Schizophrenia Bulletin, 36,* 33–35.

Van Zyl, L. E., Gaffaney, J., van der Vaart, L., Dik, B. J., & Donaldson, S. I. (2024). The critiques and criticisms of positive psychology: A systematic review. *Journal of Positive Psychology, 19,* 206–235.

Vernon, M. D. (1965). Review of the book *A short history of British psychology, 1840-1940,* by L. S. Hearnshaw. *Eugenics Review, 56,* 212–213.

Vince, G. (2013, May 17). Cities: How crowded life is changing us. *BBC Future.* https://www.bbc.com/future/article/20130516-how-city-life-is-changing-us

Vinkers, C. H., Tijdink, J. K., & Otte, W. M. (2015). Use of positive and negative words in scientific PubMed abstracts between 1974 and 2014: Retrospective analysis. *BMJ*, *351*, h6467.

Vivian, B. (2023). *Campus misinformation: The real threat to free speech in American higher education*. Oxford University Press.

Waddell, G., & Aylward, M. (2010). *Models of sickness and disability: Applied to common health problems*. Royal Society of Medicine.

Wakefield, L. (2022, September 14). LGB Alliance: Anti-trans lobby group's troubling, ugly history. *PinkNews*. http://thepinknews.com/2022/09/14/lgb-alliance-transphobia-charity-history/

Wallace-Wells, D. (2024, May 1). Are smartphones driving our teens to depression? *New York Times*. https://www.nytimes.com/2024/05/01/opinion/smartphones-social-media-mental-health-teens.html

Walsh, R. T. G., Teo, T., & Baydala, A. (2014). *A critical history and philosophy of psychology: Diversity of context, thought, and practice*. Cambridge University Press.

Wang, Q. (2016). Why should we all be cultural psychologists? Lessons from the study of social cognition. *Perspectives on Psychological Science*, *11*, 583–596.

Washington, H., Piccolo, J., Gomez-Baggethun, E., Kopnina, H., & Alberro, H. (2021). The trouble with anthropocentric hubris, with examples from conservation. *Conservation*, *1*, 285–298.

Webb, A., Heyne, G., Holmes, J. E., & Peta, J. L. (2016, April). Which box to check: Assessment norms for gender and the implications for transgender and nonbinary populations. *Division 44 Newsletter*. https://www.apadivisions.org/division-44/publications/newsletters/division/2016/04/nonbinary-populations

Weintraub, P. (2009). *Cure unknown: Inside the Lyme epidemic*. St Martin's Press.

Weiss, S. F. (2010). After the fall: Political whitewashing, professional posturing, and personal refashioning in the postwar career of Otmar Freiherr von Verscheur. *Isis*, *101*, 722–758.

Weisstein, N. (1993). Power, resistance and science: A call for a revitalized feminist psychology. *Feminism & Psychology*, *3*, 239–245.

West, P. (2024, May 8). The delusion of the pro-Palestinian campus protestors. *The Spectator*. https://www.spectator.co.uk/article/the-delusion-of-the-pro-palestinian-campus-protestors

Whittaker, A. C., Ginty, A. T., Hughes, B. M., Steptoe, A., & Lovallo, W. R. (2021). Cardiovascular stress reactivity and health: Recent questions and future directions. *Psychosomatic Medicine, 83*, 756–766.

Whittington, K. E. (2024). *You can't teach that! The battle over university classrooms.* Polity Press.

Williams, R. (1974). A history of the Association of Black Psychologists: Early formation and development. *Journal of Black Psychology, 1*, 9–24.

Williams, W. R., & Cabiles, J. J. B. (2016). Psychologists' role in contextualizing America's history of classism. *PsycCRITIQUES, 61*(50), 1.

Willingham, A. J. (2021, August 26). The most ridiculous historical arguments denying women the right to vote. *Represented by CNN.* https://edition.cnn.com/2021/08/26/us/womens-equality-day-right-to-vote-trnd/index.html

Willis, B., & Lancaster, C. L. (2020). Calling for a cease fire: Ending psychology's long conflict with religion. In L. T. Benuto, M. P. Duckworth, A. Masuda, & W. O'Donohue (Eds.), *Prejudice, stigma, privilege, and oppression: A behavioral health handbook* (pp. 447–474). Springer.

Willy, R. (1897). Die Krisis in der Psychologie. *Vierteljahrsschrift für wissenschaftliche Philosophie, 21*, 227–353.

Winkelmann, R., Donges, J. F., Smith, E. K., Milkoreit, M., Eder, C., Heitzig, J., Katsanidou, A., Wiedermann, M., Wunderling, N., & Lenton, T. M. (2022). Social tipping processes towards climate action: A conceptual framework. *Ecological Economics, 192*, 107242.

Winters, K. (2011). *Gender madness in American psychiatry: Essays from the struggle for dignity.* GID Reform Advocates.

Wittkower, E. (1946). Psychological aspects of psoriasis. *Lancet, 247*, 566–569.

Wood, N. (2020). Racism in clinical psychology within the heart of the old empire. *South African Journal of Psychology, 50*, 446–449.

Wood, N., & Patel, N. (2019). Editorial: Being nice is not enough. *Clinical Psychology Forum, 323*, 1–2.

Woolley, H. T. (1910). Psychological literature: A review of the recent literature on the psychology of sex. *Psychological Bulletin, 7*, 335–342.

Wu, K. J. (2025, January 31). CDC data are disappearing. *The Atlantic.* https://www.theatlantic.com/health/archive/2025/01/cdc-dei-scientific-data/681531/

Yakushko, O. (2019). Eugenics and its evolution in the history of western psychology: A critical archive review. *Psychotherapy and Politics International, 17*, e1495.

Yakushko, O., & Blodgett, E. (2021). Negative reflections about positive psychology: On constraining the field to a focus on happiness and personal achievement. *Journal of Humanistic Psychology*, *61*, 104–131.

Yeomans, N. D. (2011). The ulcer sleuths: The search for the cause of peptic ulcers. *Journal of Gastroenterology and Hepatology*, *26*, 35–41.

Yerkes, R. M. (1923). A foreword. In C. C. Bringham, *A study of American intelligence* (pp. v–viii). Princeton University Press.

Yip, T., Cheah, C. S. L., Kiang, L., & Hall, G. C. N. (2021). Rendered invisible: Are Asian Americans a model or a marginalized minority? *American Psychologist*, *76*, 575–581.

Yong, E. (2020, August 19). Long-haulers are redefining Covid-19. *The Atlantic*. https://www.theatlantic.com/health/archive/2020/08/long-haulers-covid-19-recognition-support-groups-symptoms/615382/

Young, J. (2009). Moral panic: Its origins in resistance, ressentiment and the translation of fantasy into reality. *British Journal of Criminology*, *49*, 4–16.

Yu, S. (2018). Uncovering the hidden impacts of inequality on mental health: A global study. *Translational Psychiatry*, *8*, 98.

Yurcaba, J. (2023, May 17). DeSantis signs 'Don't Say Gay' expansion and gender-affirming care ban. *NBC News*. https://www.nbcnews.com/nbc-out/out-politics-and-policy/desantis-signs-dont-say-gay-expansion-gender-affirming-care-ban-rcna84698

Zanghellini, A. (2020). Philosophical problems with the gender-critical feminist argument against trans inclusion. *SAGE Open*, *10*(2), 1–14.

Zimmerman, D., Attridge, J., Rolin, S., & Davis, J. (2022). Psychometric equivalence of standard and prorated Boston Naming Test scores. *Assessment*, *29*, 527–534.

GPSR Compliance
The European Union's (EU) General Product Safety Regulation (GPSR) is a set
of rules that requires consumer products to be safe and our obligations to
ensure this.

If you have any concerns about our products, you can contact us on

ProductSafety@springernature.com

In case Publisher is established outside the EU, the EU authorized
representative is:

Springer Nature Customer Service Center GmbH
Europaplatz 3
69115 Heidelberg, Germany